Polysulfide Manufacture

1970

C. Placek

Thirty-Five Dollars

NOYES DATA CORPORATION
Park Ridge, New Jersey, U.S.A.

Noyes Data, S.A.   Zug, Switzerland   London, England

FOREWORD

The detailed, descriptive information in this book is based on U.S. Patents relating to the production of polysulfides.

This book serves a double purpose in that it supplies detailed technical information and can be used as a guide to the U.S. Patent literature in this field. By indicating only information that is significant, this book then becomes an advanced review of polysulfide technology.

The U.S. Patent literature is the largest and most comprehensive collection of technical information in the world. There is more practical, commercial, timely process information assembled here than is available from any other source. The technical information obtained from a patent is extremely reliable and comprehensive; sufficient information must be included to avoid rejection for "insufficient disclosure".

The patent literature covers a substantial amount of information not available in the journal literature. The patent literature is a prime source of basic commercially utilizable information. This information is overlooked by those who rely primarily on the periodical journal literature. It is realized that there is a lag between a patent application on a new process development and the granting of a patent, but it is felt that this may roughly parallel or even anticipate the lag in putting that development into commercial practice.

Many of these patents are being utilized commercially. Whether used or not, they offer opportunities for technological transfer. Also, a major purpose of this book is to describe the number of technical possibilities available, which may open up profitable areas of research and development.

These publications are bound in paper in order to close the time gap between "manuscript" and "completed book". Industrial technology is progressing so rapidly that hard cover books do not always reflect the latest developments in a particular field, due to the longer time required to produce a hard cover book.

The Table of Contents is organized in such a way as to serve as a subject index. Other indexes by company, inventor, and patent number help in providing easily obtainable information.

CONTENTS AND SUBJECT INDEX

INTRODUCTION

Ever since polysulfide polymers (also called polysulfide rubbers in the industry and in various trades) were developed commercially, they have held a favorable position among sealants. In 1969, U.S. consumption of polysulfide polymers as sealers reached 18 million pounds out of a total sealant consumption of about 64 million pounds; only polyvinyl chloride consumption exceeded that of polysulfide polymers for this use. For the 1970's, market projections for polysulfides as sealants indicate a 5 to 6% annual increase.

Historically, commercial development of polysulfide polymers seems to have been almost exclusively in the domain of Thiokol Chemical, at least in the initial phases — thus the name Thiokol rubbers applied to this group of polymers. The basic process for the polymers was developed by J.C. Patrick and H.R. Ferguson (U.S. Patent 2,466,963) together with J.S. Jorczak and E.M. Fettes. The latter authored what has become the primary journal reference to these agents — Ind. Eng. Chem., Vol. 43, No. 2, February 1951, page 324. The Patrick and Ferguson process is the first described in this report. Thiokol's continued preeminence in this field is indicated by the number of patents described here that have been issued to that company — 32 of 73 included in this report.

The number of polysulfide polymers synthesized over the years is probably beyond counting, ranging over a wide spectrum of structures. This report, therefore, is limited to those processes dealing with polymers possessing the disulfide (—S—S—) group, one of the characteristics of the original Thiokol rubbers.

Chapter 1 concerns itself with basic processes for producing such polymers, while chapter 2 includes processes for modified polymers — those containing modifying entities ranging from arsenic to silicon. Curing methods are the primary subject of chapter 3. Chapters 4 and 5 relate to processes per se and product form, respectively, while chapter 6 discusses single-package compositions designed for use at work sites.

BASIC PROCESSES

POLYSULFIDE POLYMER — "THIOKOL RUBBER"

The rationale for the earlier processes for Thiokol rubbers was in two areas: to resolve the "cold flow" problem and to produce new products (and processes) without the necessity of using solvents or dispersion media, in a liquid form and so reactive that upon suitable treatment they may be converted into polysulfide polymers.

It had been known for a long time that in many uses of the polysulfide polymers available prior to this process, the property of "cold flow" is a disadvantage. For example, polymeric material is frequently used as a gasket to seal joints between surfaces. The gasket is placed between the surfaces, these are forced toward each other and into pressure contact with the gasket and the latter is compressed. To maintain its desired sealing effect, the gasket material should have an adequate tendency to resist the deformation caused by compression and to recover its original dimensions.

In other words, the material should possess to a considerable extent the properties of a spring, i.e., the ability of recovering its original shape after release from the action of deforming compression forces and the tendency to recover during the application of those forces.

Linear polysulfide polymers have very little power of recovery and therefore possess the disadvantage of "cold flow." By bridging the linear chains with cross connecting "links" or groups of atoms, the power of recovery may be imparted to the polysulfide polymers.

The process, developed by J.C. Patrick and H.R. Ferguson (U.S. Patent 2,466,963; April 12, 1949; assigned to Thiokol Corporation), includes improved means of rendering cross linked polysulfide polymers amenable to numerous processing treatments necessary in many uses. The process involves:

> (1) Formation of the polymers, both linear and cross-linked or three-dimensional.
> (2) Splitting or cleavage of said polymers to produce solid and liquid products and
> (3) Curing of the split products.

As an alternative to (2) synthesis of polymers having a molecular weight not exceeding 50,000 to 75,000 by oxidation of monomeric polymercaptans or lower molecular weight polythio-polymercaptans is also disclosed. This alternative may be identified as (2a).

2

Formation of the Polymers: The making of polysulfide polymers includes a number of patents issued to Patrick.

Patent No.	Date of Issue
2,049,974	August 4, 1936
2,100,351	November 30, 1937
2,142,144	January 3, 1939
2,142,145	January 3, 1939
2,195,380	March 26, 1940
2,206,641	July 2, 1940
2,206,642	July 2, 1940
2,206,643	July 2, 1940
2,216,044	September 24, 1940
2,221,650	November 12, 1940
2,235,621	March 18, 1941
2,255,228	September 9, 1941
2,278,127	March 31, 1942
2,278,128	March 31, 1942

The specific reaction is carried out by the use of 60 mols of sodium tetrasulfide containing as a dispersing agent magnesium hydroxide produced by the addition to the polysulfide solution of 160 grams of sodium hydroxide and 400 grams of crystallized magnesium chloride ($MgCl_2 \cdot 6H_2O$). The reaction mixture is heated to 60°C., and then has added to it a mixture consisting of 49 mols of dichlorodiethyl formal and 1 mol of 1,2,3-trichloropropane.

The ratio is 49 to 1 for the purpose of producing cross-linkages along the chain at an average spacing of one cross-link to each fifty units of polymer. The mixture of the organic halides is added slowly so that a period of about one hour is consumed by the addition, during which time the reacting polysulfide mixture is kept under continuous and efficient agitation to produce a highly dispersed latex-like reaction product. The cross-linked polymer produced by this reaction may be maintained in the form of a latex and utilized in this form, or may be separated by any of the known methods. In either event it may be subjected to splitting, and curing of the split or cleavage products.

Splitting or Cleavage of the Polymer: This reaction involves splitting or cleaving the polysulfide polymer at —SS— linkages thereof with generation of mercaptan or mercaptide terminals.

By using an acceptor the reactions can be made to go to any desired extent, e.g., a polymer molecule having a molecular weight of 100,000 to 200,000 may be split so that the average or statistical molecular weights of the products may be 75,000, 50,000, 40,000, 30,000, 20,000, 10,000, etc., all the way down to the monomer. The polymeric cleavage or hydrolytic products are polythiopolymercaptans which are either solid or liquid at normal temperature, e.g., 25°C. The solid products have molecular weights within the range of 15,000 to 75,000 and the liquid products have molecular weights within the range of 500 to 12,000.

Curing the Products: The coagulated or dried split polymer may be compounded with compounding and curing ingredients as follows:

	Parts by Weight
Split polymer	100
Zinc oxide	10
Zinc chromate	10
Paraformaldehyde	10
Reinforcing carbon black	60
Stearic acid	1

The above components are mixed on a conventional rubber mixing mill and placed in a steel mold and cured at a temperature of 310°F. for 30 minutes. The zinc chromate is the primary curing agent. This product, when submitted to a standard ASTM test for "cold flow" shows an 80% recovery and is in marked contrast to a product produced from a linear polymer. Extensive variations of each of these steps are possible, with a combined total of 26 being described.

OTHER BASIC THIOKOL PROCESSES

The basic Thiokol processes that quickly followed the Patrick process covered in U.S. Patent 2,466,963 include those for making polyhydroxide polysulfide polymers, polyacetal polysulfide polymers, and polysulfide polymers characterized by what is called "chemical plasticization". The first of these was developed by E.M. Fettes and the next two by J.C. Patrick.

Polyhydroxide Polysulfide Polymers

In this process, developed by E.M. Fettes (U.S. Patent 2,527,375; October 24, 1950; assigned to Thiokol Corporation), a high molecular weight polymer is formed first, and then cleaved or split to a liquid.

Formation of Polymer: To 1.5 mols of sodium disulfide add 0.08 mol of sodium hydroxide, 1 gram of a wetting agent such as sodium naphthalene sulfonate, and 0.04 mol of magnesium chloride to precipitate magnesium hydroxide in situ. Heat the reaction mixture to 70°C. and add dropwise over a period of 60 minutes a mixture of 0.8 mol of triglycol dichloride and 0.2 mol of glycerol dichlorhydrin. After all of the halogenated reactants are into the reaction vessel, the reaction should be heated to 100°C. and maintained at that temperature for a period of 30 minutes, after which the latex slurry is diluted with water and permitted to settle. The latex-like dispersion of the polymer is then washed free from soluble salts by repeated dilution with water followed by intermediate settling and dilution. The polymer thus obtained is normally solid and rubber-like when separated as a coagulum from its dispersed condition.

Conversion of Solid to Liquid Polymer: To the latex which has been washed as described above are added 0.3 mol of sodium hydrosulfide and 1.1 mols of sodium sulfite. The latex is heated with agitation in the presence of the splitting reactants (NaSH and Na_2SO_3) for about an hour at 80°C. and the solid polymer is thereby converted to one which is normally

liquid, i.e., at about 25°C. The latex is washed again as described above until free from
soluble salts and is then acidified to a pH of about 4 which causes the breaking of the latex
and separation of the viscous liquid. This viscous liquid can then be dried by any suitable method.

The product consists of a mixture of polymers having an average molecular weight much
lower than that of the polymer from which the product was made. The molecules of the
product are characterized by (a) disulfide linkages, (b) mercaptan terminals and (c) hydroxyl groups along the polymeric chains. They are therefore polythiopolyhydroxy polymercaptans. The polymer made by the procedure (a) was split at —SS— linkages according
to the reaction

$$---RSSR--- + NaSH \longrightarrow ---RSNa + HSR--- + S$$
$$S + Na_2SO_3 \longrightarrow Na_2S_2O_3$$

RSS represents an average copolymeric unit.

The NaSH is the splitting reagent and the Na_2SO_3 is a sulfur acceptor which disturbs the
equilibrium and enables the splitting reaction to go forward. In general, splitting may be
effected by a compound M_2P where P is oxygen or sulfur and M is an alkali metal, ammonium or hydrogen, in the presence of a compound which accepts P and combines therewith
to form a stable, nonoxidizing compound. For example, water is a splitting agent in the
presence of an oxygen acceptor, e.g., nascent hydrogen.

The dichlorhydrin used was a mixture of 1,2-dichloro-3-propanol and 1,3-dichloro-2-pro-
panol in the ratio of about 75% of the former and 25% of the latter. Instead of this specific
product, any of the functional hydroxyl compounds set forth in the table may be substituted
in the same mol proportion and, instead of the triglycol dichloride, organic compounds in
general having a halogen atom or other replaceable substituent attached to each of two
different carbon atoms may be employed.

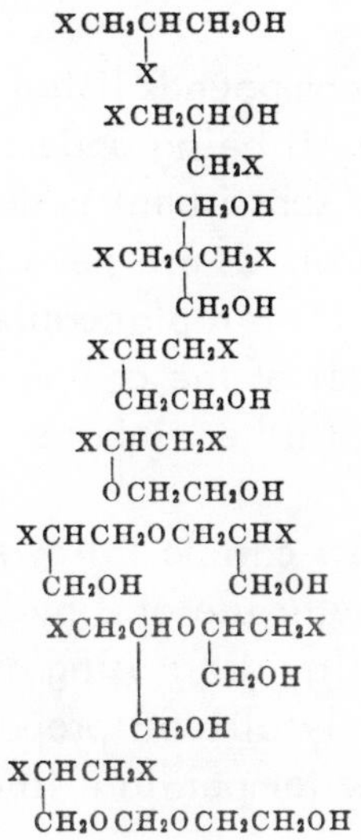

$$XCH_2CHCH_2OH$$
$$X$$
$$XCH_2CHOH$$
$$CH_2X$$
$$CH_2OH$$
$$XCH_2CCH_2X$$
$$CH_2OH$$
$$XCHCH_2X$$
$$CH_2CH_2OH$$
$$XCHCH_2X$$
$$OCH_2CH_2OH$$
$$XCHCH_2OCH_2CHX$$
$$CH_2OH \quad CH_2OH$$
$$XCH_2CHOCHCH_2X$$
$$CH_2OH$$
$$CH_2OH$$
$$XCHCH_2X$$
$$CH_2OCH_2OCH_2CH_2OH$$

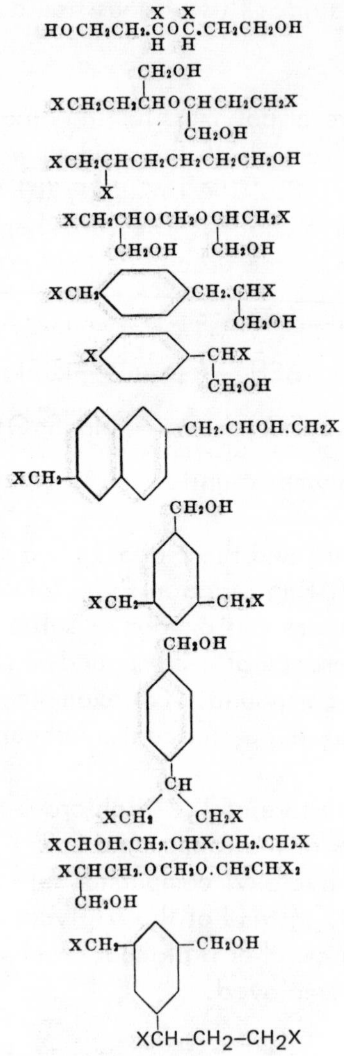

$$HOCH_2CH_2.C\underset{H}{\overset{X}{C}}O\underset{H}{\overset{X}{C}}.CH_2CH_2OH$$

$$\underset{\underset{CH_2OH}{|}}{XCH_2CH_2.CHOCHCH_2CH_2X}$$
(with CH$_2$OH above)

$$XCH_2CHCH_2.CH_2.CH_2.CH_2OH$$
(with X below)

$$\underset{\underset{CH_2OH}{|}}{XCH_2CHOCH_2OCHCH_2X}\underset{CH_2OH}{|}$$

Structures with ring:

XCH$_2$ —〈ring〉— CH$_2$.CHX
   with CH$_2$OH

X —〈ring〉— CHX
   with CH$_2$OH

XCH$_2$ —〈fused rings〉— CH$_2$.CHOH.CH$_2$X

Ring with CH$_2$OH (top), XCH$_2$ (left), CH$_2$X (right):

CH$_2$OH
XCH$_2$ —〈ring〉— CH$_2$X

Ring with CH$_2$OH top and CH below:

CH$_2$OH
〈ring〉
CH

XCH$_2$ CH$_2$X

$$XCHOH.CH_2.CHX.CH_2.CH_2X$$
$$\underset{\underset{CH_2OH}{|}}{XCHCH_2.OCH_2O.CH_2CHX_2}$$

Ring with XCH$_2$ (left), CH$_2$OH (right):

XCH$_2$ —〈ring〉— CH$_2$OH

XCH—CH$_2$—CH$_2$X

For example, any of the numerous compounds listed in U.S. Patent 2,216,044 may be substituted for the triglycol dichloride, it being understood as explained in that patent that where halogen or other replaceable substituent is directly connected to the carbon of an aromatic nucleus that higher temperatures are necessary to split off that replaceable substituent than when one is dealing with a replaceable substituent on a methylene carbon atom. It will be noted that a number of the compounds listed in U.S. Patent 2,216,044 are characterized by the presence of ether linkage.

A number of variations of the process can be followed, for example, use of a monofunctional compound to limit chain growth; use of a hydroxy compound exclusively and using a monofunctional halide to obtain liquidity; using an excess of the functional organic compound in relation to the alkaline polysulfides; preparation of copolymer using mercapto oxidation; and use of relatively low temperature to control the molecular size of the product.

The polymers are characterized by the presence of polymeric units having the formula SRS (which may also be written RSS) where R is a radical selected from the group consisting of

$$-\overset{\displaystyle |}{\underset{\displaystyle |}{C}}-\overset{\displaystyle |}{\underset{\displaystyle |}{C}}-$$

designating adjacent carbon atoms and

$$-\overset{\displaystyle |}{\underset{\displaystyle |}{C}}\ldots\overset{\displaystyle |}{\underset{\displaystyle |}{C}}-$$

designating carbon atoms joined to and separated by intervening structure and that at least some of said units contain an alcohol group. While in many cases the polymer will be composed largely or wholly of units having the above identified skeleton carbon structure, the polymer need not always be composed exclusively of such units since some of the units may have a skeleton carbon structure symbolized by the expression

$$-\overset{\displaystyle |}{\underset{\displaystyle |}{C}}-$$

where C is a single carbon atom. For example, a copolymer may be made by reacting an alkaline sulfide with a mixture of alpha, beta-dichlorhydrin, 1,5-dichlorpentane and methylene dichloride. The skeleton carbon structure of the units of which the resulting polymer is composed will be

$$-\overset{|}{\underset{|}{C}}-\overset{|}{\underset{|}{C}}-\overset{|}{\underset{|}{C}}, \quad -\overset{|}{\underset{|}{C}}-\overset{|}{\underset{|}{C}}-\overset{|}{\underset{|}{C}}-\overset{|}{\underset{|}{C}}-\overset{|}{\underset{|}{C}}-$$

and

$$-\overset{\displaystyle |}{\underset{\displaystyle |}{C}}-$$

Instead of methylene dichloride other compounds having two halogens (or equivalent replaceable groups) attached to the same carbon atom may be used, e.g., benzal chloride. Another identifying characteristic of the polymers is the fact that upon exhaustive treatment with an alkaline hydrosulfide in conjunction with an alkaline sulfite, cleavage occurs with the production of one or more polymercaptans containing at least two —SH groups and at least one hydroxyl group.

The polymers may be succinctly described as polyhydroxy polythio polymers and also as polyhydroxy polythio polymercaptans existing normally, i.e., at 25°C., in a liquid condition. The structure may be described more in detail (in addition or alternatively to the description previously given) by stating that those polymers comprise or include units containing at least two carbon atoms, that some of the units may contain only one carbon atom, that the units of which the polymer is composed are connected together by sulfur linkages at least some of which are disulfide (—SS—) linkages, that at least some of said units contain at least one alcohol group and that upon exhaustive treatment of said polymer with an alkaline hydrosulfide in conjunction with an alkaline sulfite, one or more monomeric polymercaptans are obtained containing at least one alcohol group.

Polyacetal Polysulfide Polymers

A process developed by J.C. Patrick (U.S. Patent 2,527,376; October 24, 1950; assigned to Thiokol Corporation) is described in the following examples.

Example 1:   Two mols of 2,3-dichloro propanol-1 are dissolved in about 200 cubic centimeters of benzene and slightly more than 1 mol of formaldehyde, preferably in the form of paraformaldehyde, is added. A trace of acid catalyst is put into the mixture such as, for example, one drop of concentrated hydrochloric acid, and the mixture is refluxed at the boiling point of benzene in an esterification flask fitted with a trap for the removal of water and connected so that the benzene will be continuously returned to the reaction flask. The refluxing with removal of water is continued until substantially one mol of 18 cubic centimeters of water have been removed. The mixture in the flask is then transferred to a regular distilling apparatus and the benzene is distilled off, leaving the reaction product which is the tetrachloro dipropanol formal. The product obtained is a slightly viscous liquid boiling at 145° to 148°C. at 3 mm. pressure and having a specific gravity of 1.35.

Example 2:   Proceed as above, substituting 1,3-dichloro propanol-2 for the 2,3-dichloro propanol-1. The product is a white crystalline solid, melting at 53°C. and boiling at 130° to 132°C. at 2 mm. pressure.

Example 2A: Proceed as in either of Examples 1 or 2, substituting a mixture of the chloro compounds for the single compounds. The product is a liquid.

Example 3:   Two mols of allyl alcohol are treated with slightly more than one mol of formaldehyde in the presence of a trace of acid, e.g., HCl, as a catalyst to give diallyl formal. The ethylenic linkages are then saturated with a halogen to give a tetrahalo compound corresponding to the product formed. The preferred method of performing this reaction is to dissolve the diallyl formal in CCl_4 and add the halogen in solution in CCl_4 to the solution of diallyl formal compounds.

Example 4:   Proceed as in Example 1 or 2, substituting alpha-monochlorohydrin for the 2,3-dichloro propanol-1 and slightly more than 2 mols of formaldehyde. A polymer is formed which is a viscous amber liquid, specific gravity 1.26. By reacting for a longer time, a wax is obtained.

Example 5:   Proceed as in Example 4, substituting beta-monochlorhydrin for the alpha hydrin. The product obtained is similar to that of Example 4.

Example 6:   Three liters of a 2-molar solution of sodium disulfide are placed in a three-necked flask of five liter capacity equipped with means for mechanical agitation and a thermometer to indicate temperatures. To this solution is added a solution of 10 grams of NaOH in 15 cc of water followed by a solution of 25 grams of crystallized magnesium chloride ($MgCl_2 \cdot 6H_2O$) in 50 cc of water. The mixture is heated at a temperature of about 160°F., and to it are added a mixture of 5 mols of dichlorodiethyl formal and 0.025 mols of either of the compounds or the equivalent proportions of the polymers obtained above. The addition of the mixed chlorides is carried out slowly in such a manner that the complete addition requires about one hour.

During the addition of the mixed chlorides an exothermic condition takes place, and the temperature rises to about 180°F. When the reaction is completed, the temperature is maintained at 180°F. for approximately one hour, after which the latex that has been formed as a result of the reaction is permitted to settle out of the reaction liquid and the supernatant liquid is then removed by decantation or siphoning.

The latex is treated with successive washes of warm water until entirely free from water-soluble impurities, after which it is transferred to a suitable receptacle and dilute acid is added until the reaction of the supernatant liquid is brought to a pH of about 6, whereupon a phenomenon analogous to coagulation of rubber latex takes place. The coagulum so-formed is then kneaded with cold water until every trace of residual acid is removed, after which the soft elastic mass is dried.

In the above reaction of Example 6, various soluble sulfides may be substituted for the sodium disulfide; in fact, soluble sulfides in general may be so substituted, e.g., soluble sulfides having the formula MS_{2-6} where M is an alkali or alkaline earth metal or ammonium or substituted ammonium. Moreover, the corresponding monosulfides may be also substituted. Various mixtures of monosulfides and polysulfides may also be used.

Polymer Characterized by Chemical Plasticization

The other process developed by J.C. Patrick (U.S. Patent 2,553,206; May 15, 1951; assigned to Thiokol Corporation) gives a polymer which is characterized by what may be termed "chemical plasticization," i.e., a polymer, the plasticity of which is obtained and controlled not by the physical admixture of a plasticizer but by providing atomic structure which is an integral part of the molecules of the polymer and which controls the plastic qualities within desired limits.

The process may be applied to produce a polysulfide polymeric product substantially odorless and tasteless. Such a polymer has a number of important uses in applications where freedom from odor and/or taste is necessary, e.g., as a sealing, coating or lining for food containers. In addition, the process gives a polymer which may be successfully used instead of or in conjunction with natural chicle in the manufacture of chewing gum or chewing gum bases. A combination of conditions is observed to realize the maximum advantages of the process:

    (1) An organic dihalogen compound is employed having linear chains of at least six atoms between the halogen atoms. The atoms of said chain are selected from the group consisting of carbon, oxygen and sulfur atoms and at least five of the atoms of said chain are carbon atoms.

    (2) The halogen compound is reacted with an alkaline or ionizable sulfide reagent which may be symbolized by the expression $MS_{1.1-1.9}$ where S is a sulfur atom and M is an inorganic or organic cation illustrated by alkali metals, alkaline earth metals, ammonium, alkyl ammonium, alkanol radicals, etc.

The use of this reagent introduces monosulfide as well as disulfide linkages into the chains of the polymer, creates the internal or structural plasticization, and makes it possible to carry out step (3) described below without rendering the polymer too tough or intractable for many purposes. The structural or "chemical" plasticization mentioned has a great advantage over the use of physically ad-mixed plasticizers because the latter have a certain amount of vapor pressure tending to impart odor and/or taste and in numerous instances even introduce toxicity into the product.

(3) The treatment with the sulfide reagent is carried out so as to convert into high polymers any low polymers in the product produced as in step (2) or to solubilize said low polymers whereby they are readily removed. This treatment involves subjecting the polymer obtained as above described in step (2) to treatment with an alkaline sulfide. The proportion of sulfur in the alkaline sulfide used in this treatment should be not less than that indicated by the formula $MS_{1.1-1.9}$. If a sulfide of this character is used in this step (3), steps (2) and (3) can be combined. Preferably, however, the sulfide used in step (3) is a disulfide or other polysulfide in which the sulfur rank is about 2 or between about 2 and 6, and in this case, step (3) is separate and distinct from step (2). In said step (3) it is also possible to use an alkaline monosulfide.

Example 1: Use of an alkaline sulfide having a composition indicated approximately by the formula $MS_{1.1-1.9}$ in the initial reaction is followed by an after treatment involving use of a sulfide having a composition within the range indicated by the formula MS_{2-6}.

Into a flask fitted with condenser and means for mechanical agitation and of not less than 3 liters capacity are introduced 2,000 cc of a 2-molar solution having a composition indicated by the empirical formula $Na_2S_{1.5}$. To the sulfide solution are added about one gram of a suitable wetting agent which may be, for example, a sodium naphthalene sulfonate and 6 grams of freshly precipitated magnesium hydroxide. This mixture is heated with agitation to a temperature of about 60°C. and 3 mols of dichloro diethyl formal are added to the heated polysulfide mix at such a rate that about 30 minutes are required for the complete introduction of the organic halide. Since this reaction is somewhat exothermic the temperature is controlled by any suitable means in such a way that it does not go above a temperature of about 90°C. during the introduction of the organic halide.

After all of the organic halide has been added to the polysulfide mix the temperature is raised to 100° to 105°C. for about 30 minutes. Water is added to dilute the dispersion to the capacity of the reaction vessel, the agitator is stopped and the latex-like polymer dispersion is allowed to settle out. The supernatant liquid is withdrawn by any suitable means such as decantation and the latex is washed once with water and allowed to settle again thereby removing most of the soluble by-products of the reaction. The supernatant liquid is again withdrawn and 2,000 cc of a molar solution of Na_2S_2 are added to the concentrated latex dispersion. The reaction vessel is heated again with agitation to a temperature of about 90°C. for about 30 minutes, after which the latex is washed free from water-soluble materials.

The washed latex-like dispersion procured as described above may be used in the dispersed form or if so desired it can be coagulated to a soft rubber-like mass by acidification with any suitable acid, for example, acetic acid, or dilute hydrochloric acid, until the pH of the mixture after acidification is about 4. This treatment causes the coagulation of the dispersed polymer. The polymer so obtained is substantially odorless and tasteless and will withstand exacting tests in this respect. For example, it may be heated to a temperature of about 300°F. without developing any disagreeable odor. It will be understood, of course, that in the initial reaction expressed above instead of an alkaline sulfide having the formula $MS_{1.5}$, any alkaline or ionizable sulfide having a composition expressed approximately by the formula $MS_{1.1-1.9}$ may be used in the initial reaction and that in the after treatment instead of using an alkaline disulfide, any alkaline, ionizable or water-soluble polysulfide having a composition expressed approximately by the formula MS_{2-6} may be employed.

Example 2: Modified procedure in which instead of using an after treatment a single treatment with an alkaline or ionizable sulfide is employed, the proportion thereof being in considerable excess and the composition thereof being expressed approximately by the formula $MS_{1.1-1.9}$.

Proceed as in Example 1 except that 4,000 cc of a 2-molar solution of a sulfide having the empirical formula $Na_2S_{1.4}$ is substituted for the 2,000 cc of $Na_2S_{1.5}$ described in Example 1. After all the dihalide is added to the flask, the temperature in this case is raised to 100° to 105°C. for 60 minutes instead of 30 minutes after which the polymer dispersion is treated exactly as described in Example 1.

Example 3: Proceed as in Example 2 using a portion of the sulfide reagent say about half in the initial reaction and the remainder as an after treatment.

Example 4: To use an alkaline monosulfide for the after treatment, proceed as in Example 1 except that in the after treatment 1,000 cc of a 2-molar solution of sodium monosulfide are used instead of the 1,000 cc of 2-molar disulfide shown in Example 1, and the heat treatment after the addition of the monosulfide is changed to 60 minutes instead of the 90 minutes shown in Example 1, after which the polymer dispersion is treated exactly as in Example 1.

POLYMERIZATION OF DISULFIDE RING COMPOUNDS

A process developed by F.O. Davis (U.S. Patent 2,657,198; October 27, 1953; assigned to Reconstruction Finance Corporation) deals with compounds having the generic formula

$$S(CH_2)_{1\ to\ 3}Z(CH_2)_{1\ to\ 3}S$$

their production and polymerization. The general formula of a preferred class of the cyclic disulfides is

$$SC_2H_4ZC_2H_4S$$

Z is a member of the group consisting of $-O-$, $-S-$, $-OCH_2O-$, $-SCH_2S-$, $-OC_2H_4O-$, $-SC_2H_4S-$, and $-CH_2-$.

The monomeric ring compounds are polymerized by treatment with a catalyst. The polymerization of these disulfide rings is quite surprising and would not be expected from the chemical structure of the compounds. The organic aliphatic disulfides, either monomeric or polymeric, are quite stable compounds. The monomers have the following structures:

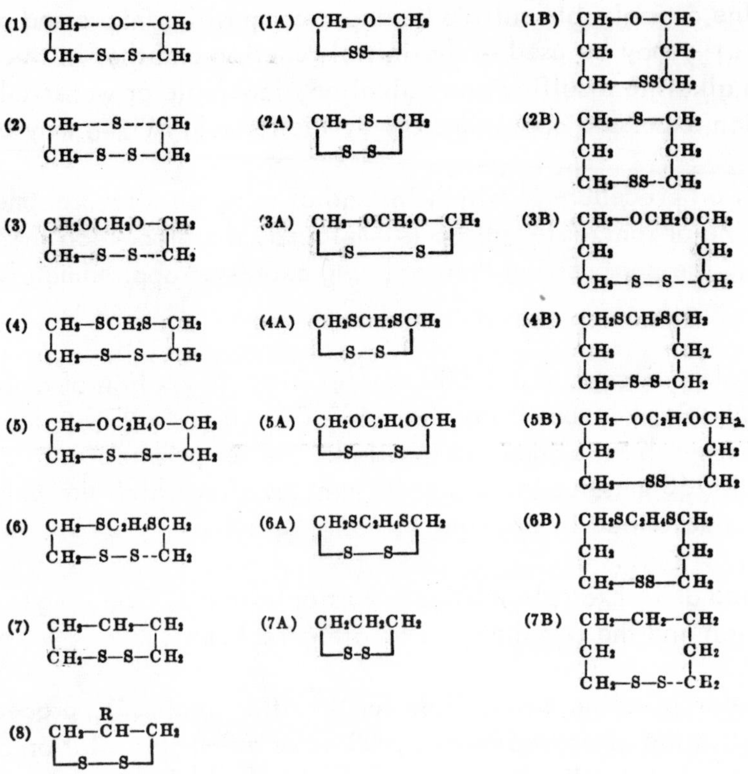

(1)
$$CH_2---O---CH_2$$
$$CH_2-S-S-CH_2$$

(1A)
$$CH_2-O-CH_2$$
$$\lfloor\;\;SS\;\;\rfloor$$

(1B)
$$CH_2-O-CH_2$$
$$CH_2\;\;\;\;\;CH_2$$
$$CH_2---SSCH_2$$

(2)
$$CH_2---S---CH_2$$
$$CH_2-S-S-CH_2$$

(2A)
$$CH_2-S-CH_2$$
$$\lfloor\;\;S-S\;\;\rfloor$$

(2B)
$$CH_2-S-CH_2$$
$$CH_2\;\;\;\;\;CH_2$$
$$CH_2-SS--CH_2$$

(3)
$$CH_2OCH_2O-CH_2$$
$$CH_2-S-S---CH_2$$

(3A)
$$CH_2-OCH_2O-CH_2$$
$$\lfloor\;\;S\;\;\;\;S\;\;\rfloor$$

(3B)
$$CH_2-OCH_2OCH_2$$
$$CH_2\;\;\;\;\;\;\;CH_2$$
$$CH_2-S-S--CH_2$$

(4)
$$CH_2-SCH_2S-CH_2$$
$$CH_2-S-S-CH_2$$

(4A)
$$CH_2SCH_2SCH_2$$
$$\lfloor\;\;S-S\;\;\rfloor$$

(4B)
$$CH_2SCH_2SCH_2$$
$$CH_2\;\;\;\;\;\;CH_2$$
$$CH_2-S-S-CH_2$$

(5)
$$CH_2-OC_2H_4O-CH_2$$
$$CH_2-S-S---CH_2$$

(5A)
$$CH_2OC_2H_4OCH_2$$
$$\lfloor\;\;S\;\;\;\;\;S\;\;\rfloor$$

(5B)
$$CH_2-OC_2H_4OCH_2$$
$$CH_2\;\;\;\;\;\;\;\;CH_2$$
$$CH_2-SS---CH_2$$

(6)
$$CH_2-SC_2H_4SCH_2$$
$$CH_2-S-S--CH_2$$

(6A)
$$CH_2SC_2H_4SCH_2$$
$$\lfloor\;\;S\;\;\;\;\;S\;\;\rfloor$$

(6B)
$$CH_2SC_2H_4SCH_2$$
$$CH_2\;\;\;\;\;\;\;\;CH_2$$
$$CH_2-SS--CH_2$$

(7)
$$CH_2-CH_2-CH_2$$
$$CH_2-S-S-CH_2$$

(7A)
$$CH_2CH_2CH_2$$
$$\lfloor\;\;S-S\;\;\rfloor$$

(7B)
$$CH_2-CH_2-CH_2$$
$$CH_2\;\;\;\;\;\;CH_2$$
$$CH_2-S-S--CH_2$$

(8)
$$CH_3-\overset{R}{CH}-CH_3$$
$$\lfloor\;\;S\;\;\;\;S\;\;\rfloor$$

Example: The preparation of Compound 1 follows. 3 mols, 972 cc of 3.09 molar $Na_2S_{4.27}$, was treated with one gram of the sodium salt of butyl naphthalene sulfonic acid, eight grams of sodium hydroxide and 25 grams of $MgCl_2 \cdot 6H_2O$ all used as approximately 25% solutions. This reaction mixture was heated to a temperature of 140°F. and there was added to this reaction mix 2.7 mols (386 grams) of $ClC_2H_4OC_2H_4Cl$.

The feed period was 90 minutes during which a latex formed in the reaction. This latex was distilled with steam until 1,000 cc of distillate had been collected to remove all of the congeneric 1,4-thioxane formed in the reaction. After one washing the latex was treated with 4.5 mols (180 grams) sodium hydroxide for one hour at 180°F. The latex was washed free of polysulfide solution and subjected to steam distillation. The distillate in this case was cloudy with droplets of oil settling out.

Decantation gave approximately 1 1/2 grams of oil per 500 cc of distillate but this amount could be somewhat increased by extraction with ethyl ether. Although the rate of formation of the oil, as evidenced by its rate of removal in the steam distillation, was quite slow, it continued practically unchanged for a considerable period of time, for example, about two months of distillation. If other halides are used other polymers are obtained which give different cyclic materials having different physical properties and different degrees of stability. To produce Compound 1A, proceed as above using dichloro methyl ether instead of bis(beta-chloroethyl) ether. To produce Compound 1B proceed as above using bis(gamma-chloropropyl) ether. The other structures are produced by essentially the same method, with appropriate changes in the starting materials.

Compounds 1, 1A, and 1B are polymerized in the following manner. 100 cc of the cyclic dithiodiethyl oxide (Compound 1 above), which is an oil at room temperature, are mixed at room temperature, e.g., 25°C. with 2 cc of a 25% solution of sodium methylate in methanol and the mixture is poured into a mold. Polymerization begins immediately without heating and in about 24 hours a rubbery polymer is obtained which becomes tough in about 48 hours. Heat is evolved during the reaction.

With 10 cc of the sodium methylate solution, the polymerization proceeds faster and a tough rubber is obtained within a few minutes. The other cyclic disulfides are polymerized in much the same way. To polymerize two different monomers with each other, for example, Compounds 1 and 2, the polymerization is done as with Compounds 1, 1A, and 1B using the sodium methylate solution (above).

50 cc of Compound 1 and 50 cc of Compound 2 are used with 10 cc of the sodium methylate solution. In this case a fairly rapid polymerization takes place. The product is a copolymer having rubbery properties which is composed of equimolecular proportions of thioether and ether disulfides. Latices of these types, if mixed, would not give copolymers unless given a polysulfide treatment.

A variety of catalysts may be used including alkali alcoholates and alkaline sulfides, hydrosulfides and polysulfides, e.g., the sulfides, hydrosulfides and polysulfides of sodium, potassium, ammonium, calcium, barium etc. Even water alone acts as a catalyst. Other classes of materials which can be effectively used as catalysts are those of the group consisting of alkyl and aralkyl-substituted ammonia compounds and alkyl and aralkyl-substituted ammonium compounds.

LOW CHLORINE CONSUMPTION

A process developed by F.O. Davis (U.S. Patent 2,728,748; December 27, 1955; assigned to Thiokol Chemical Corporation) produces polysulfide polymers with a greatly reduced consumption of chlorine.

The polymers are made in an intermediate form with a very low chlorine consumption and having a relatively low molecular weight. Those intermediate polymers can then be converted into high molecular weight polymers by further condensation with no additional chlorine consumption. As shown in Case A, a polymer having a molecular weight of about 1,800 is obtained with a chlorine consumption of only 2 chlorine atoms, as compared with a chlorine consumption of 22 chlorine atoms to obtain the same product by the conventional reaction of 11 mols of dichlorodiethyl formal and 10 mols of Na_2S_2. These polymers have the general formula $(FR''S)_2 + (n-2)_y (SRS)_x (R'S_n)_y$ where F is a member of the group consisting of hydroxyl and halogen, n varies from 2 to 6, x varies from 2 to 200 and y varies from 0 to 100, R and R' being oxahydrocarbon aliphatic radicals, R having a sulfur connected valence of only 2 and R' having a sulfur connected valence of n, and R" being a radical of the group consisting of alkylene radicals and bivalent aliphatic oxahydrocarbon radicals.

In the process, a dithiodiglycol having the general formula $HOR'''SSR'''OH$, for example dithiodiethylene glycol, is reacted with formaldehyde and a dihalogenated formal having the general formula $ClR'''OCH_2OR'''Cl$, as for example beta,beta'-dichlorodiethyl formal, where R''' is an alkylene radical having one to six carbon atoms.

By controlling the ratio of formaldehyde to glycol, polymers can be obtained having either chlorine terminals or hydroxyl terminals, and by controlling the mol ratio of dichlor formal to the glycol the degree of polymerization can be controlled. The reaction can be symbolized as follows: Case A symbolizes the production of chlorine terminated polymers, and Case B, hydroxyl terminated polymers.

Case A and Case B describe the reaction as occurring in several steps to indicate more clearly the probable mechanism of the reaction. However, the reaction may be conveniently carried out in a single stage where all of the several steps occur substantially simultaneously.

Case A: Chlorine terminated polymers — In each case, purely for purposes of illustration dithiodiethylene glycol will be taken to illustrate the glycol and beta,beta'-dichlorodiethyl formal will be taken to illustrate the halogenated formal employed.

1. $10HOC_2H_4SSC_2H_4OH + 9CH_2O \rightleftharpoons$

$$HO(C_2H_4SSC_2H_4OCH_2O)_9C_2H_4SSC_2H_4OH + 9H_2O$$

2.   Step 1 Polymer + $ClC_2H_4OCH_2OC_2H_4Cl \rightleftharpoons$

$$HO(C_2H_4SSC_2H_4OCH_2O)_9C_2H_4SSC_2H_4OCH_2OC_2H_4Cl + ClC_2H_4OH$$

3.   Step 2 Products + $CH_2O \rightleftharpoons$

$$ClC_2H_4OCH_2O(C_2H_4SSC_2H_4OCH_2O)_9(C_2H_4SSC_2H_4OCH_2O)C_2H_4Cl + H_2O$$

Case B: Hydroxyl terminated polymers —

1. $9HOC_2H_4SSC_2H_4OH + 8CH_2O \rightleftharpoons$

$$HOC_2H_4S(SC_2H_4OCH_2OC_2H_4S)_8SC_2H_4OH$$

2.   Step 1 Product + $ClC_2H_4OCH_2OC_2H_4Cl \rightleftharpoons$

$$HOC_2H_4S(SC_2H_4OCH_2OC_2H_4S)_8SC_2H_4OCH_2OC_2H_4Cl + ClC_2H_4OH$$

3.   Step 2 Product + $HOC_2H_4SSC_2H_4OH \rightleftharpoons$

$$HOC_2H_4S(SC_2H_4OCH_2OC_2H_4S)_8SC_2H_4OCH_2OC_2H_4SSC_2H_4OH + ClC_2H_4OH$$

This product may be written as

$$HOC_2H_4S(SC_2H_4OCH_2OC_2H_4S)_9SC_2H_4OH$$

The formula of the product produced in Case A is

(1) $ClR'''OCH_2O(R'''SSR'''OCH_2O)_aR'''Cl$ identical with

(2)

$$\underset{\substack{|\\R''}}{ClR'''OCH_2OR'''}\overset{}{S}\underset{\substack{|\\R}}{(SR'''OCH_2OR'''S)}_{a-1}\underset{\substack{|\\R''}}{SR'''OCH_2OR'''Cl}$$

when $a-1 = x = 9$.

The reaction is carried out by refluxing the reactants in the presence of a water-immiscible solvent boiling at 80° to 140°C. in reflux apparatus provided with a trap which separates the water-solvent mixture, returns the solvent automatically to the reaction mixture and enables the water to be separated. The reaction is continued until water ceases to be evolved. In the production of the hydroxyl terminated polymers (Case B) ethylene chlorhydrin or other alkylene chlorhydrin, e.g., $ClR'''OH$, will be evolved with and pass into the water condensate and may be recovered therefrom. Typical solvents which may be employed for the purpose indicated include benzene, toluene, xylene, ethylene dichloride and propylene dichloride.

In each instance the reactants are refluxed in the presence of the solvent, as stated, water is collected and separated by means of the trap, and the refluxing is continued until water ceases to be evolved. The solvent is then distilled off and the desired polymer remains as a residue. The chlorinated formal splits off a small proportion of hydrochloric acid which supplies a catalyst and it is not necessary to add an additional catalyst. However, from 0.1 to 1% by weight, based on the glycol $HOR'''SSR'''OH$, of a catalyst may be added. Typical catalysts are hydrochloric acid, sulfuric acid, phosphoric acid, toluene sulfonic acid and ferric chloride.

LIQUID POLYTHIOPOLYMERCAPTANS

While it is possible to make a liquid polymer composed substantially wholly of linear polymeric chains and cure it to a solid rubbery material, the product thus produced possesses properties that are undesirable for most applications. Thus the cured linear polymers are susceptible to "cold flow," that is, they undergo permanent deformation when subjected to pressure at atmospheric temperatures. Also they lack resilience and strength.

To achieve a product having the desired strength, resilience and freedom from cold flow it is desirable to introduce a moderate amount of cross-linking into the polymer. Cross-linking is effected by utilizing organic radicals which are connected to sulfur atoms of more than two of the disulfide linkages, that is to say, RS_3 units and RS_4 units are interposed at intervals in polymeric chains that are essentially composed of SRS units.

A very large number of organic radicals R can be used in the preparation of the polysulfide polymers. Examples of useful radicals are given in U.S. Patents 2,142,144; 2,142,145;

2,195,380; 2,306,643; 2,216,044; 2,221,650; 2,235,621; 2,278,127; 2,363,114 and 2,363,615. However, while a large number of different types of organic radicals can be used in the preparation of the polysulfide polymers, the radicals that are of primary commercial interest fall in a somewhat more limited group and for the most part are saturated aliphatic hydrocarbon or oxahydrocarbon radicals. Specific examples of useful organic radicals within this group are given hereinafter.

Until this process, developed by E.M. Fettes and E.R. Bertozzi (U.S. Patent 2,875,182; February 24, 1959; assigned to Thiokol Chemical Corporation), most of the commercial liquid polysulfide polymers have been made by the "splitting" method disclosed in U.S. Patent 2,466,963 above.

In a typical example of this method an organic halide or mixture of halides, quite commonly a mixture containing a major proportion of dihalide and minor proportion of trihalide, is reacted with an alkaline polysulfide to form a high molecular weight solid polymer and is then split with a suitable splitting agent, e.g., sodium hydrosulfide, and a sulfur acceptor, e.g., sodium sulfite, to produce a material of lower molecular weight. When the splitting reaction is properly carried out as disclosed in U.S. Patent 2,466,963, a liquid polymer having mercaptan terminals is produced that can be effectively cured to produce a rubbery solid having the desired characteristics. However, it is apparent that this method involves a rather round-about route for obtaining a liquid polymer from low molecular weight raw materials and that it would be desirable to have a simpler, more direct method of preparing the liquid polymer.

This process provides a relatively simple and effective method of making liquid polysulfide polymers. Also, it gives a method of making polysulfide polymers with improved yield by both decreasing losses due to the formation of soluble mercaptides and reducing the formation of undesirable by-products. Moreover, it is a direct method of making liquid polysulfide polymers from relatively inexpensive raw materials. It provides a method of making polysulfide polymers that permits a high degree of control over the molecular weight of the product.

The process consists of reacting one or more organic halides with an aqueous inorganic thiosulfate solution to form an organic thiosulfate and thereafter treating the organic thiosulfate solution with a mixture of a water-soluble hydrosulfide and sulfide to produce the desired polymer. The initial reaction between the organic halide and the thiosulfate may be represented by the following equation:

$$(1) \qquad RX_n + nNa_2S_2O_3 \longrightarrow R(S_2O_3Na)_n + nNaX$$

In this equation n has a value of 2 to 4, X is halogen and R as indicated above may be any of a large variety of organic groups, but for economic reasons is usually selected from the group consisting of aliphatic hydrocarbons and oxahydrocarbon radicals. It is usually desirable that RX_n be composed primarily of RX_2 compounds with relatively small proportions of RX_3 and/or RX_4 compounds being included to provide the desired cross-linking in the finished polymer as described above. It will of course be understood that other water-soluble metal thiosulfates can be substituted for the sodium thiosulfate specifically referred to in the above equation.

In the second step of the process one or more organic thiosulfates prepared in the manner indicated in Equation (1) are reacted with an alkali metal monosulfide and an alkali metal hydrosulfide in accordance with the following equation:

$$(2) \quad zR'(S_2O_3Na)_2 + yR''(S_2O_3Na)_m + (z-1+y)Na_2S + [2+(m-2)y]\,NaSH \longrightarrow$$

$$H_{[2+(m-2)y]}(SR'\text{-}S)_z(R''S_m)_y + (z-1+y)Na_2SO_3 + [z+1-y(1-m)]Na_2S_2O_3$$

In this equation m is 3 or 4, y may vary from 0 to 10, z may vary from 2 to 500 and R' and R'' are selected from the group consisting of aliphatic hydrocarbon and oxahydrocarbon radicals, R' having two sulfur-connected valences and R'' having m sulfur-connected valences. Thus the expression SR'S represents the divalent units of which the polymeric chains are essentially composed and the expression $R''S_m$ represents the units that provide cross-linking between the polymeric chains.

In the foregoing equation sodium monosulfide and hydrosulfide are illustrative of the reagents that may be used. Water-soluble hydrosulfides and sulfides of other metals can be substituted for the sodium salts of the equation. Also, sulfides of higher rank than the monosulfide, e.g., sulfides having a rank of 1 to 6, may be used.

Example 1:   A 12-liter flask was charged with 7,600 ml. of water, 4,960 grams (20 mols) of $Na_2S_2O_3 \cdot 5H_2O$ and 34 grams (0.32 mol) of Na_2CO_3. The mixture was heated with agitation to 200°F., then 1,144 grams (8.0 mols) dichloroethyl ether was added and heating was continued at 200°F. with agitation for 3 hours to produce a dithiosulfate of ethyl ether. Titration of unreacted thiosulfate at the end of this period indicated a 96% conversion of the chloroether to the thiosulfate.

This solution was treated with a mixture of sodium monosulfide and sodium hydrosulfide in the following manner: A 5-liter flask was charged with 700 ml. H_2O, 208 grams of flake sodium monosulfide containing 60% by weight (1.6 mols) Na_2S, and 64 grams of flake sodium hydrosulfide containing 70% by weight (0.80 mol) NaSH. The mixture was heated to 180°F. with agitation. Thereafter 2,875 ml. of the organic thiosulfate solution as prepared above and containing about 2.0 mols of the organic thiosulfate was run into the mixed sulfide solution at a uniform rate over a period of an hour. The reaction mixture was agitated at a temperature of about 180°F. during addition of the thiosulfate and for another hour after the addition of the organic thiosulfate was complete. A liquid polysulfide polymer formed during the reaction period and settled to the bottom of the reaction flask.

The aqueous phase in the reaction flask was separated by decantation and the liquid polymer was washed three times by decantation for about 5 minutes with fresh hot water at a temperature of 140° to 160°F. Separation of the wash water was effected by permitting the polymer to settle and decanting the water. The polymer was then oven-dried overnight at 180° to 200°F. and filtered. A yield of 184 grams, 68% of the theoretical yield, was obtained. The viscosity of the product polymer was 21 poises.

The liquid polymer was cured in the following manner: 100 grams of the liquid polymer was mixed on a paint mill with 50 grams of titanium dioxide (Titanox T102), 7 grams of p-quinone dioxine (GMF), 3 grams diphenylguanidine and 0.2 gram of sulfur. The mixture was caused to make three passes through the paint mill, then placed in an oven for 16 hours at 158°F. The cured sample was molded into sheets in a press at 250°F. for ten minutes. The physical properties of the resulting sheets as determined on a Scott tester and Shore durometer were as follows:

Ultimate tensile strength	200 psi
Maximum elongation	900 %
Shore A durometer hardness	28

Example 2:   A cross-linked polymer was made by copolymerizing thiosulfates derived from dichloroethyl ether and from trichloropropane. To prepare the ethyl ether thiosulfate a reactor was charged with 31 gallons of water, 167.7 lbs. $Na_2S_2O_3 \cdot 5H_2O$, 38.9 lbs. of dichloroethyl ether and 450 grams of sodium carbonate. The resulting mixture was heated to and maintained at 200°F. for a period of 5 hours to produce the ethylether dithiosulfate. Titration of unreacted thiosulfate at the end of the reaction period showed 95.2% conversion of the dichloroethyl ether to the organic thiosulfate.

To prepare the propane-1,2,3-tris-thiosulfate a second reactor was charged with 750 ml. water, 465 grams (1.875 mols) of $Na_2S_2O_3 \cdot 5H_2O$, 4 grams (0.04 mol) of sodium carbonate and 250 grams of the methyl ether of diethylene glycol. This mixture was heated to 200°F. with agitation and 74 grams (0.50 mol) of 1,2,3-trichloropropane was rapidly added with agitation. Heating of the mixture was continued at reflux temperature (216° to 222°F.) with agitation for 7 hours. The course of the reaction was followed by titration of unreacted thiosulfate with iodine. The conversion rose gradually to 86% at the end of the 7-hour period, by which time the pH had dropped to about 5 and the reaction was considered complete.

The solutions containing the thiosulfate derived from dichloroethyl ether and the thiosulfate derived from trichloropropane respectively were then mixed and the organic thiosulfates were copolymerized. A flask was charged with 700 ml. of water, 708 grams of flake sodium sulfide containing 60% (1.6 mols) of Na_2S, and 64 grams of flake sodium hydrosulfide containing 70% (0.8 mol) of NaSH. The resulting solution was heated to and maintained at 180°F. with agitation. 2,810 ml. of the ethyl ether thiosulfate solution containing 1.96 mols on an ether basis was mixed with 122 ml. (0.04 mol on a trichloropropane basis) of the organic thiosulfate derived from trichloropropane, and the resulting mixture was fed to the mixed sulfide solution according to the procedure of Example 1 to form a liquid polysulfide polymer that was cross-linked.

This polymer was compounded, cured and tested according to the same procedure as above and gave the following results.

Ultimate tensile strength	250 psi
Maximum elongation	280 %
Shore A durometer hardness	44

Several variations of the above methods are possible.

LIQUIDS BY SPLITTING OF HIGH POLYSULFIDE POLYMERS

Polysulfide polymers may be divided broadly into two classes (a) the normally solid polymers having molecular weights within the range of from 15,000 to 200,000 or greater and (b) the normally liquid polymers having molecular weights within the range of from 500 to 12,000. The normally solid or high polysulfide polymers may be linear, partially cross-linked or completely cross-linked. The liquid polymers can be obtained by splitting the high polymers into lower polymers.

J.C. Jorczak and E.M. Fettes in Ind. and Eng. Chem., vol. 43, pp. 324-328 (1951), make polysulfide liquid polymers by reductive cleavage of disulfide groups in high polysulfide polymers to yield products which have terminal thiol groups. The procedure employed is to treat a water dispersion of a polysulfide polymer with sodium hydrosulfide and sodium sulfite. The sodium hydrosulfide splits a disulfide link to form a thiol and a sodium salt of a thiol, and depending upon the mol ratio of sodium hydrosulfide to polymer repeating segments, liquid polymers of varying molecular weights can readily be prepared. The disadvantages of carrying out the reductive cleavage of disulfide groups in high polysulfide polymers in a water dispersion and recovering the liquid polymers are apparent.

A process developed by W.R. Nummy (U.S. Patent 2,919,262; December 29, 1959; assigned to The Dow Chemical Company) shows that normally solid polysulfide polymers of high molecular weight can readily be split or converted to polymers of lower molecular weight having reactive thiol or mercapto terminal groups by treating the high polysulfide polymers with polythiols, e.g., dithiols, trithiols, tetrathiols, etc.

The polythiols employed in effecting the reductive cleavage of the disulfide groups in the high polysulfide polymers to yield polysulfide liquid polymers enter into and become an integral part of the liquid product or the rubbery polysulfide polymers obtained by subsequent curing or oxidation of the liquid product. The process provides an improved method for making polysulfide liquid polymers from high molecular weight solid polysulfide polymers.

Any polythiol containing two or more mercapto groups in the molecule, e.g., dithiols, trithiols, tetrathiols, etc., can be employed to effect reductive cleavage of disulfide groups in high polysulfide polymers to yield products which have terminal thiol groups. The polythiols are preferably aliphatic thiols containing from 2 to 4 thiol groups in the molecule such as polythiols having the general formula $HSR(SH)_x$ wherein R is a radical selected from the group consisting of

$$\begin{matrix} | & | \\ -C-C- \\ | & | \end{matrix} \quad \text{and} \quad \begin{matrix} | & | \\ -C\ldots\ldots C- \\ | & | \end{matrix}$$

designating respectively adjacent carbon atoms and carbon atoms joined to and separated by intervening structure and x is a whole number from 1 to 3. Suitable dithiols having the above mentioned general formula are as follows:

$$HSCH_2CH_2SH$$

$$HSCH_2CH_2CH_2SH$$

(continued)

$$HSCH_2CHCH_3$$
$$|$$
$$SH$$

$$HSCH_2CH_2CH_2CH_2SH$$

$$HSCH_2CH_2CH_2CH_2CH_2CH_2SH$$

$$HSCH_2CH_2OCH_2CH_2SH$$

$$HSCH_2CH_2OCH_2CH_2OCH_2CH_2SH$$

$$HSCH_2CH_2OCH_2CH_2OCH_2CH_2OCH_2CH_2SH$$

$$HSCH_2CH_2OCH_2CH_2OCH_2CH_2OCH_2CH_2OCH_2CH_2SH$$

$$HSCH_2CH_2OCH_2CH_2OCH_2CH_2OCH_2CH_2OCH_2CH_2OCH_2CH_2SH$$

$$HSCH_2CH_2OCH_2CHOCH_2CH_2SH$$
$$|$$
$$CH_3$$

$$HSCH_2CH_2OCH_2CH_2CH_2OCH_2CH_2SH$$

$$HSCH_2CH_2OCH_2CH_2CH_2CH_2OCH_2CH_2SH$$

$$HSCH_2CH_2OCH_2CHOCH_2CH_2SH$$
$$|$$
$$CH_2CH_3$$

Suitable tri- and tetramercapto compounds are as follows:

$$HSCH_2CHCH_2SH$$
$$|$$
$$SH$$

$$HSCH_2CH_2OCH_2CHCH_2OCH_2CH_2SH$$
$$|$$
$$OCH_2CH_2SH$$

$$HSCH_2CHCH_2OCH_2OCH_2CH_2SH$$
$$|$$
$$SH$$

$$CH_2SH$$
$$|$$
$$HSCH_2CCH_2SH$$
$$|$$
$$CH_2SH$$

$$HSCH_2CHCH_2OCH_2CHCH_2SH$$
$$|\qquad\qquad|$$
$$SH\qquad\qquad SH$$

$$HSCH_2CHCH_2OCH_2OCH_2CHCH_2SH$$
$$|\qquad\qquad\qquad|$$
$$SH\qquad\qquad\qquad SH$$

Among the polythiols having the above general formula the polythiols having from 2 to 3 mercapto groups in the molecule are preferred. Such polythiols include tetramethylenedithiol-1,4; hexamethylenedithiol-1,6; 1,2,3-trimercaptopropane; 2,2'-oxy-dipropanethiol;

1,2,3-oxy-tripropanethiol; 1,3-oxydipropanethiol; and dithiols having the general formula:

$$HSCH_2CH_2O(C_nH_{2n}O)_mCH_2CH_2SH$$

where n and m independently represent an integer from 1 to 4. The polythiols having the above general formula can be employed to split or effect the reductive cleavage of disulfide groups in any of the polysulfide polymers of high molecular weight to yield products which have terminal thiol groups. The higher molecular weight polysulfide polymers can be those prepared by the reaction of organic dihalides and sodium polysulfide such as are described in Ind. and Eng. Chem., vol. 28, pp. 1145-1149 (1936), or polysulfide polymers of high molecular weight prepared by the oxidation of one or more polythiols having the above general formula.

In the preparation of polysulfide liquid polymers, the normally solid polysulfide polymers are mixed with one or more of the polythiols in proportions corresponding to from 2.5 to 100 parts by weight of the polythiol per 100 parts of the solid polysulfide polymer. The reaction which occurs readily at room temperature or thereabout can be carried out at temperatures between 20° and 220°C., preferably from 20° to 120°C., and at atmospheric or substantially atmospheric pressure. The reaction is usually carried out with stirring of the ingredients in admixture with one another and with limited access of air or oxygen by way of a reflux condenser to the reactants, or carrying out of the reaction under an atmosphere of an inert gas, e.g., nitrogen or helium, until a uniform liquid composition is obtained. Upon conversion of the mixture of starting materials to a smooth syrup or uniform liquid, the resulting polysulfide liquid polymer is suitable for use without further treatment or purification.

The molecular weight of the polysulfide liquid polymers can be varied over a wide range depending for the most upon the relative proportions of the solid high molecular weight polysulfide polymer and the polythiol or polythiols employed. An increase in proportion of the polythiol relative to the polysulfide to be reacted therewith results in a decrease in the average molecular weight of the product.

In general when employing a solid polysulfide polymer such as that prepared by reaction of dichlorodiethyl formal and sodium sulfide as starting material and 2,2-oxybis-(ethyleneoxy)diethanethiol in proportions of from 2.5 to 100 parts by weight of the dithiol per 100 parts by weight of the polysulfide polymer, liquid products having absolute viscosities between 500 and 100,000 centipoises at 25°C. are obtained. By employing a mixture of a dithiol and a tri- or tetrathiol, polysulfide liquid polymers are obtained which are capable of being cured or oxidized to form cross-linked polysulfide polymers that are insoluble in usual organic solvents for linear polysulfide polymers.

The polysulfide liquid polymers can all be cured or vulcanized by suitable treatment with oxidizing agents such as oxygen, organic peroxides, metallic oxides, etc., in admixture with a small amount, e.g., from 1 to 10% by weight, of a tri- or tetrathiol as cross-linking agent, to form tough rubbery compositions which are suitable for a variety of applications.

POLYMERS FROM DITHIOALCOHOLS AND THIONYL CHLORIDE

A process developed by B.D. Stone (U.S. Patent 3,053,816; September 11, 1962; assigned to Monsanto Chemical Company) reacts substantially equimolar proportions of an alkanedithiol compound having from 4 to 6 carbon atoms with the thiol or mercaptan radicals attached to the terminal carbons thereof with thionyl chloride. The polymer obtained has an empirical chemical constitution of approximately $(C_xH_{2x}S_{2.5})_n$ where x is a whole number of from 4 to 6 and n represents the number of units in the polymer. Neither the structure nor the exact molecular weight of the polymer are definitely known. However, it is believed that the polymer has a molecular arrangement of the type

$$[-S-(CH_2)_x-S-S-S-(CH_2)_x-S-]_n$$

where x and n are as defined above. From the evidence available it is believed that n has a value on the order of 10 to 100, giving the polymeric material molecular weight on the order of 3,000 to 30,000.

The obtaining of this rubber-like polymer by this method is critical in that rubbery polymers are not obtained when these alkanedithiols are reacted with other sulfur chlorides, such as sulfur dichloride and sulfuryl chloride. Similarly, rubbery polymers are not obtained when alkanedithols of lower carbon content, such as 1,2-ethanedithiol, are reacted with thionyl chloride. In all of those cases, hard crumbly waxes are obtained. The reaction of this process proved to be more complex and to yield a polymer completely different in physical properties from polymers obtained from other sulfur chlorides.

The thionyl chloride and the 4 to 6 carbon alkanedithiol reactant can be contacted in any known manner. For example, the thionyl chloride can be added to the alkanedithiol reactant, or the alkanedithiol reactant can be added to the thionyl chloride. Also the reactants may be added simultaneously to the reaction vessel. In any event, the reaction is preferably carried out in the presence of an inert diluent, such as chloroform, ethyl ether, dioxane, alkylene chlorides, e.g., methylene chloride, hydrocarbons, such as benzene, toluene, etc. However, such inert diluents or solvents are not essential to the conduct of the reaction.

Reaction between the thionyl chloride and the alkanedithiol materials can be conducted at ordinary, decreased, or elevated temperatures. Temperatures of -20° to 80°C. can readily be used with temperatures of from 0° to 80°C. being preferred.

When the thionyl chloride and 4 to 6 carbon alkanedithiol are combined, the initial reaction is exothermic and continues to evolve heat until one-half of the thionyl chloride has been consumed. Up to this point, only HCl is evolved and sulfur from the thionyl chloride stays in the polymer. After half of the reaction is completed, e.g., when one-half of the molar quantities of thionyl chloride has been reacted, the reaction is no longer exothermic and both HCl and a sulfur-containing gas are evolved.

Example: 28.65 grams of 1,5-pentanedithiol (0.2102 mols) was dissolved in 170 ml. of distilled chloroform and the solution placed in a 4-necked 500 ml. flask equipped with a stirrer, reflux condenser, thermometer, and dropping funnel. The condenser exit led to two bubbler bottles in series each containing 30 g. of sodium hydroxide in 100 ml. of water.

The system was flushed with nitrogen and then 24.75 g. (0.208 mol) of freshly distilled thionyl chloride in 70 ml. of chloroform was added through the dropping funnel into the pentanedithiol solution while maintaining the temperature of the reaction mixture at about 10°C. The first half of the thionyl chloride was added over 30 minutes. Thereafter the heat evolution subsided and the remainder of the thionyl chloride was added in 5 minutes.

The reaction mixture was stirred for 1 hour at 0° to 5°C., then heated at reflux temperature for an hour and a half. Finally the chloroform was distilled off and the polymer was heated under vacuum to 75°C. for 30 minutes. Then chloroform was added again to dissolve the polymer and the resulting solution was poured into excess hexane using good agitation to precipitate the polymer. The polymer was dried overnight in a vacuum desiccator. The rubber-like polymer product weighed 26.4 g.

	Found	Calculated for $C_5H_{10}S_{2.5}$
Percent C	40.19	40.00
Percent H	6.74	6.66
Percent S	52.47	53.33

XYLYLENE POLYSULFIDE AND ALKYLENE POLYSULFIDES

A process developed by K.C. Tsou (U.S. Patent 3,054,781; September 18, 1962; assigned to The Borden Company) gives a polymer of xylylene polysulfide which has a softening point of around 250°C., as compared to only about 125°C. for the corresponding product from ethylene. It provides also a process of manufacture that gives polysulfides of the alkylenes, such as ethylene, that are substantially free of or greatly reduced in odor as compared to the commercial polysulfides on the market.

The process is illustrated by the reaction below, when ethylene dibromide and sodium thiosulfate are used as the original reactants and hydrogen peroxide at a pH below 7 as the condensing agent.

$$Br-CH_2CH_2-Br + 2Na_2S_2O_3 \longrightarrow NaO_3S_2-CH_2-CH_2-S_2O_3Na + 2NaBr$$

$$2nNaO_3S_2-CH_2-CH_2-S_2O_3Na + 2nH_2O_2 \xrightarrow{H^+}$$
$$(-S-CH_2CH_2-S-S-CH_2CH_2-S)_n + 4nNaHSO_4$$

The n is an integral number greater than 1.

Example 1: The preparation of the sodium salt of xylylene dithiosulfate is as follows. Sodium thiosulfate in amount corresponding to 63.2 parts on the anhydrous basis (approximately 0.4 mol) was dissolved in a mixture of 200 parts of water and 30 parts of ethanol. Into this there was stirred 35 parts of alpha,alpha'-dichloroxylylene (2.2 mol), the xylylene being para. The mixture was allowed to come to reflux and was then cooled as required to avoid violent boiling. When the exothermic reaction had been substantially completed as shown by subsidence of the refluxing and dropping of the temperature, the product so-formed was the intermediate sodium salt of dithiosulfate of xylylene

$$CH_2S_2O_3Na$$

$$CH_2S_2O_3Na$$

The di(monosodiumthiosulfate) so-formed remains dissolved. It may be recovered in solid form by evaporating the solution. The thiosulfate group is probably predominantly or wholly in the form of the monosodium salt

$$-S-\overset{\overset{O}{\|}}{\underset{\underset{O}{\|}}{S}}-ONa$$

Example 2: The preparation of xylylene polysulfide from xylylene dithiosulfate would then proceed in the following manner. The solution of the xylylene dithiosulfate made as the final product of Example 1 and being in solution was cooled to 0°C. and acidified with 0.5 part, more or less, of concentrated sulfuric acid which was stirred in slowly in amount to establish a pH of about 2.

Next there was introduced, gradually and with stirring, 51.5 parts of aqueous hydrogen peroxide solution of peroxide concentration 35% (1 mol of H_2O_2). The mixture was allowed to rise slowly to approximately room temperature, at which point an exothermic reaction appeared. This reaction caused the mixture to reflux vigorously and a white precipitate to settle, the boiling being moderated with cooling until the refluxing ceased and the temperature began to fall. The white precipitate was filtered, washed with a small proportion of water and then dried at a moderately elevated temperature below the melting point of any solid material present. The precipitate was xylylene polydisulfide which may be written as $(-SCH_2-C_6H_4-CH_2S-SCH_2-C_6H_4-CH_2S)_n$. This product is insoluble in all common organic solvents such as alcohol, acetone, benzene, chloroform, dimethyl formamide and dimethyl acetamide.

This product was then mixed dry with elemental sulfur in the proportion of 1 atom to each atom of combined sulfur. The mix was warmed until the whole fused and until no further substantial change in consistency occurred and the added sulfur was combined, as at 160° to 180°C. The sulfurs in the formula above are now united severally to an additional sulfur, thus

$$-\overset{\overset{}{S}}{\underset{\underset{S}{\|}}{}}-$$

The product is now xylylene poly-polysulfide, a tough rubbery material.

Example 3: The preparation of sodium salt of ethylene dithiosulfate is as follows. Sodium thiosulfate, in the amount of 99.2 parts of the anhydrous material (0.62 mol), was dissolved in 200 parts of water. To this solution was added with stirring a solution of 37.6 parts of ethylene dibromide (0.2 mol), in 33 parts of ethanol.

The mixture was allowed to reflux until homogeneous, was then cooled to 0°C. and condensed and then treated with additional sulfur in the manner described in Example 1. The resulting polymer of ethylene polysulfide had very little odor.

Example 4: The preparation of copolymers of polyxylylene and polyethylene disulfide is as follows. Equimolecular proportions of the sodium salt of xylylene dithiosulfate and the sodium salt of ethylene dithiosulfate, prepared as in Examples 1 and 3, respectively, are mixed together and the mixture condensed in the same manner as above, by 35% hydrogen peroxide after acidification. A white precipitate was collected. The structure of the resulting mixed disulfide is:

$$(-S-CH_2CH_2-S-S-CH_2-\langle\bigcirc\rangle-CH_2-S-)_x$$

When this disulfide copolymer is heated with sulfur in amount providing 1 elemental S for each combined S, the poly-polysulfide is formed. It is a tan colored, rubbery material.

POLY(METHYLENE SULFIDES)

Solid resinous products are obtained by saturating 37% aqueous formaldehyde with hydrogen sulfide at 40° to 50°C. A product obtained this way has been described as melting at 80°C. and containing 51.5% sulfur. If instead of using hydrogen sulfide an alkali metal sulfide or polysulfide is used, the products are then said to be "poly(methylene sulfides)." These polymers are described on page 190 in J.F. Walker's "Formaldehyde," second edition, Reinhold Publishing Corp., New York (1953), as ranging from hard, brittle solids to elastomers and are said to conform to:

$$HS-CH_2S-CH_2-S...CH_2-SH$$

and

$$HS-CH_2\overset{S}{\overset{\|}{S}}-CH_2-\overset{S}{\overset{\|}{S}}...CH_2-SH$$

The product from formaldehyde and sodium sulfide is a colorless amorphous powder, which becomes plastic and rubbery on heating and is insoluble in organic solvents. The elastomeric products corresponding to $(-CH_2-S_2-)_x$ are said to harden, become brittle, and lose sulfur on storage. From formaldehyde and a sodium polysulfide corresponding to the tetra or pentasulfide, rubber-like materials which show little or no change on storage are said to be obtained. A high molecular weight polymethylene sulfide is said to be obtained from hydrogen sulfide and hexamethylenetetramine. However, from this reaction, the only products which have been obtained are said to be mixtures of trithiane and materials containing nitrogen.

In none of the above work has a high molecular weight thermoplastic, stable, high-melting poly(methylene sulfide) been described but a process developed by J. Harmon (U.S. Patent 3,070,580; December 25, 1962; assigned to E.I. du Pont de Nemours and Company) gives such a product.

The process consists of self-condensing methanedithiol, $CH_2(SH)_2$, or mercaptomethyl sulfide, $HSCH_2SCH_2SH$, in the presence of a basic material at 25° to 250°C., and maintaining these conditions until essentially no more hydrogen sulfide is evolved from the reaction mixture.

In one method for preparing the poly(methylene sulfides), a reactor is charged with methanedithiol or mercaptomethyl sulfide. The reactor is swept with nitrogen and there is added at ambient temperature (ca. 25°C.) under nitrogen from 0.05 to 15% of a basic catalyst, e.g., ammonia or a trialkylamine, based on the weight of the methanedithiol or mercaptomethyl sulfide. After effervescence, due to the liberation of hydrogen sulfide, has subsided, the solution is heated gradually to between 140° and 200°C. and held at this temperature until essentially no more hydrogen sulfide continues to be released which, depending upon temperature and catalyst used, may require from 15 minutes to 30 hours. The reaction mixture is allowed to cool and the reactor is opened.

The reaction mixture is extracted with a suitable solvent, e.g., chloroform or benzene, then with hot water, and dried under vacuum. The resulting product, which generally melts above 220°C., may be treated with a strong mineral acid, e.g., aqueous hydrochloric acid at 90° to 100°C. After removal of acid, the polymer is heated in the solid state in the presence of an alkaline material such as ammonia, at temperatures approaching the melting point for several hours. The polymer produced thereby generally melts above 240°C. shows a very high degree of crystallinity, and is readily melt-spinnable.

The acid treatment step is unnecessary in certain cases where the entire procedure has been carried out in a completely inert atmosphere. The product of this process is a thermoplastic, high molecular weight, crystalline poly(methylene sulfide) characterized by being fiber-forming, by having strong x-ray diffraction lines corresponding to interplanar spacings of 4.370 to 4.43 angstrom units, 2.985 to 3.015 angstrom units, and 2.17 angstrom units, and by having a melting point of at least 220°C.

POLYSULFIDE RUBBERS FROM $C_{12}H_{24}X_2$ DIHALIDES

According to a process developed by M.B. Berenbaum, N.A. Rosenthal and H.Z. Lecher (U.S. Patent 3,099,643; July 30, 1963; assigned to Thiokol Chemical Corporation), contrary to what would be expected, polysulfide polymers can be prepared from dihalides of the formula $C_{12}H_{24}X_2$ which have good resiliency and excellent low temperature properties.

More particularly, these polymers have glass transition temperature values (T_g) within the range −110° to −120°F. and are capable of retaining their resiliency when kept at −40°F. for periods of as long as five months. It has been further found that such polymers are stable at relatively high temperatures and that thermal decomposition of such polymers occurs only above 250° to 260°C. Since these results were obtained without any compounding of the polymer, presumably even better results can be obtained by compounding and addition of suitable modifying agents to such polymers. The starting material for preparation of the polymers may be conveniently prepared from 1,3-butadiene by a recently available process. In accordance with the disclosure of U.S. Patent 2,850,538, saturated C_{12} glycols can be prepared by dimerizing butadiene in the presence of sodium and a dialkyl ether, reacting the resulting product with ethylene oxide, hydrolyzing the reaction mixture to form C_{12} unsaturated glycol, and hydrogenating to produce saturated glycols. The glycols thus produced can be converted to omega-omega dihalides by any of various known processes including reaction with hydrohalogen acids, followed by distillation to remove excess acid and water.

The foregoing process normally produces a mixture of three isomers. Thus a typical mixture resulting from this process may comprise 30% by weight of a linear omega-omega C_{12} dihalide, 55% by weight of an ethyl-substituted C_{10} omega-omega dihalide, and 15% by weight of a diethyl-substituted C_8 omega-omega dihalide. While such a mixture may be used as such as a starting material for preparation of the polymers, it has been found that better properties are achieved if only the branched chain dihalides are used. Thus it is preferable to remove by recrystallization the straight chain C_{12} dihalide from the mixture and employ as a starting material the resulting mixture of the ethyl C_{10} and diethyl C_8 dihalides.

In general, the polymers can be prepared by methods similar to those previously used in preparing polysulfide polymers from lower molecular weight dichlorides, i.e., by reacting the dichlorides with aqueous alkali metal polysulfides. However, it has been found necessary to carry out the reaction at higher temperatures and pressures than those previously used, i.e., elevated pressures of up to 500 psi and temperatures of 100° to 200°C.

Example: An aqueous solution of $Na_2S_{2.25}$ was prepared having a concentration such that 530 ml. of the solution contained 1.2 mols of polysulfide. An autoclave equipped with a mechanical stirrer, thermometer, and dropping feed arrangement, was charged with 530 ml. of this solution and 265 ml. of methanol. There was then added to the autoclave 7.5 ml. of Nekal solution (5% aqueous alkyl naphthalene sulfonic acid), 3.1 ml. of NaOH solution (50% concentration), and 21.8 ml. of $MgCl_2$ solution (25% concentration) to produce a magnesium hydroxide dispersion in the polysulfide solution. This mixture was heated to 102°C. and the pressure in the autoclave adjusted to 200 psi.

The dichloride feed comprised a mixture of ethyl C_{10} omega-omega dichloride and diethyl C_8 omega-omega dichloride in a ratio of about 55:15. A mixture consisting of 0.7 mol (166 grams) of this dichloride and 0.014 mol (2.1 grams) of 1,2,3-trichloropropane was prepared and added dropwise to the polysulfide solution over a period of thirty minutes. The temperature of the reaction mixture was then brought to 120°C. and maintained at this temperature for about four hours to produce a polysulfide polymer in latex form. The latex was removed from autoclave and washed free of excess polysulfide with water.

The product thus obtained was still soft and so was toughened according to the following procedure: The latex was reintroduced into the autoclave with 725 ml. of 2.35 molar $Na_2S_{2.25}$ and heated at 160°C. under 200 psi pressure for four hours. The resulting latex was washed free of polysulfide and then subjected to a second toughening treatment. In the second treatment the latex was mixed with 1486 ml. of $Na_2S_{2.25}$ containing 2.4 mols of the polysulfide and heated at 180°C. and 200 psi for an additional four hours. The resulting latex was washed free of polysulfide with water and coagulated. The coagulated product had a hardness of 25 as measured on the Shore A durometer at room temperature.

A sample of this material was put in a refrigerated container which was set to hold a constant temperature of –40°C. From time to time the hardness of the samples was determined using a Shore A–Z durometer #4518. The results are given in the following table.

Hours at –40°C.	Hardness
3	44
144	44
600	47
1166	45
2630	44
3590	46

The foregoing data show that there was no significant change in the hardness of this material at a temperature of –40°C. over a period of about four months.

POLYSULFIDE POLYMERS VIA ORGANOMETALLICS AND SULFUR HALIDES

In a process developed by R.E. Robinson and M.W. Mueller (U.S. Patent 3,119,795; January 28, 1964; assigned to National Distillers and Chemical Corporation), polysulfide polymers may be prepared in good quality and high yields by reacting a dialkali metal hydrocarbon with a sulfur halide. Suitable for use as the dialkali metal hydrocarbon are substances such as disodiooctadiene, isomeric mixtures thereof, and, in general, dialkali metal hydrocarbons of 4 to 20 carbon atoms. Sulfur chlorides such as sulfur monochloride and sulfur dichloride are preferred, but other halides may be used.

The quantity of dialkali metal hydrocarbon employed may range from about 0.5 to about 2 mols per mol of sulfur halide. However, the reaction is carried out using approximately a mol to mol ratio of the dialkali metal hydrocarbon to sulfur halide.

When used, the amount of reaction medium is not critical, but generally is used in an amount corresponding to about 1 to 100 parts of diluent to 1 part of sulfur halide, and preferably about 5 to 70 parts of diluent to 1 part of sulfur halide. The reaction of dialkali metal hydrocarbon with a sulfur halide takes place readily at any temperature below the decomposition temperature of the particular reactants employed. For example, when substances such as disodiooctadiene and sulfur monochloride are used, the reaction may occur at a temperature between approximately –75° and +75°C. The polysulfide polymers produced possess utility as fuel binding agents in solid propellants for rockets, in oil- and solvent-resistant gaskets, in sealing compounds, and the like.

Example: An oven-dried flask was charged with 7 parts (0.05 mol) of sulfur monochloride dissolved in 200 parts (by volume) of n-hexane. The flask was fitted with a paddle-type stirrer, a thermometer, an outlet to a nitrogen bubbler system, and a tube for dialkali metal hydrocarbon addition. To the addition tube was added 50 parts (0.003 mol) of 0.6 molar disodiooctadiene in alkylate diluted with 50 parts of n-hexane. The disodiooctadiene was added over a period of about 110 minutes while the temperature was held at about 25° to 30°C.

The mixture was then stirred for about 20 minutes and allowed to stand overnight. It was then treated with 200 parts of water and filtered by suction, giving 6 parts (84% of theoretical) of an amorphous solid which softened slightly at 175°C. and decomposed at 205°C. The solid was compressed in a Carver press at 175°C. to yield a dark, smooth film of rubber. The product had the following composition: 46.6% C, 5.8% H, 38.1% S, and 9.5% Cl. Changes in various parameters, amount of reagent, temperature, etc. give polymers with varying compositions and properties.

POLYMER MADE FROM $C_2S_3Cl_2$

A process developed by S. Proskow (U.S. Patent 3,234,187; February 8, 1966; assigned to E.I. du Pont de Nemours and Company) gives a polymer made by reacting the compound $C_2S_3Cl_2$ with either (1) an alkali metal sulfide, (2) an alkali metal trithiocarbonate or (3) hydrogen sulfide in the presence of a hydrogen chloride acceptor. Although the $C_2S_3Cl_2$ used as a starting material in the immediate process has been described in the literature, its precise structure is unknown. Evidence points to the dichloromethylene trithiocarbonate structure

$$Cl_2C \overset{\displaystyle S}{\underset{\displaystyle S}{<>}} C=S$$

The preparation of the compound is given by M. Delepine et al., Bull. Soc. Chim. France (5), 2, 1969 (1935). It can also be made as follows: A mixture of 115 g. of tetrachloro-1,3-dithiethane, 77.5 ml. of thiolacetic acid, and 500 ml. of dry diethyl ether was stirred at 25°C. After 0.5 hour, 77.5 ml. of dry dimethylformamide was added and the resulting solution stirred at 25°C. for 17 hours. The solution gradually turned red-brown and a black, viscous layer separated. The ether layer was decanted, washed successively with cold water, dilute hydrochloric acid, and water, and then dried over anhydrous magnesium sulfate.

Evaporation of the ether on the steam bath gave a viscous, liquid residue. This residue was dissolved in 300 ml. of carbon tetrachloride and the solution stirred at 25°C. while 25 ml. of bromine was added dropwise. The precipitated bromine adduct was filtered, washed with carbon tetrachloride, and dried in vacuo. The bromine adduct was covered with 200 ml. of acetone and the mixture heated to boiling in the presence of decolorizing charcoal. After filtration and cooling to -80°C., the crystals were filtered, redissolved in fresh acetone (200 ml.), and recrystallized by cooling to -80°C. Filtration under nitrogen, followed by brief drying with nitrogen and in air between pressed filters gave 48 g. of pure $C_2S_3Cl_2$ as bright yellow crystals melting at 53° to 54°C. In one revision of the process, a solution in absolute ethanol of the chlorocarbon sulfide, $C_2S_3Cl_2$, is placed in a reactor and cooled to 0°C. or lower. To the cooled solution there is then added with stirring a solution of an alkali metal trithiocarbonate (or alkali metal sulfide) and the mixture allowed to warm to ambient temperature. The product separates as a pale tan solid. This solid is collected, washed successively with hydrochloric acid, acetone, and carbon bisulfide, and then dried in vacuo.

This polymer melts at about 200°C., exhibits broad infrared absorption bands at about 8.84μ, 945μ, and 14.0μ, yields self-supporting films, is resistant to degradation by air and by boiling 10% aqueous hydrochloric acid, consists essentially of carbon and sulfur in an atomic ratio of about 1:2, and has a molecular weight in excess of 800.

In a modification of the procedure, $C_2S_3Cl_2$ dissolved in a solvent, e.g., dimethylformamide, is cooled to 0°C., stirred, and to the stirred solution hydrogen sulfide is added continuously. After turbidity appears, the solution is allowed to warm to ambient temperature, while addition of hydrogen sulfide is continued. The pale tan solid which separates is collected and processed as previously described. In this modification, the presence of a hydrogen chloride, or acid, acceptor is essential also. The dimethylformamide functions both as a solvent and an acid acceptor.

Example: A solution of 1.91 g. of $C_2S_3Cl_2$ in 20 ml. of dry dimethylformamide, under substantially anhydrous conditions, was stirred at 0°C. and hydrogen sulfide was then slowly bubbled into the solution. After the solution became turbid, i.e., in about 10 minutes, it was removed from the cooling bath and hydrogen sulfide was then bubbled in for about 2.5 additional hours. The polymer which formed was collected by filtration and washed by stirring successively for 0.5 hour with dilute hydrochloric acid, acetone, and carbon bisulfide. The washed product was dried in vacuo at 56°C. There was obtained 0.4 g. of a pale tan solid which melted at 195° to 200°C. The product analyzed 14.99% C and 81.10% S.

Improved yields of polymer were obtained by carrying out the reaction at lower temperatures and employing triethylamine as the hydrogen chloride acceptor:

A solution of 1.91 g. of $C_2S_3Cl_2$ in 50 ml. of ethylene glycol dimethyl ether was cooled to -80°C., and gaseous hydrogen sulfide was passed into the solution for 10 minutes. The resulting solution was then stirred and maintained at -80°C. while a solution of 1.0 g. of triethylamine in 10 ml. of ethylene glycol dimethyl ether was added in small portions. After the addition, the mixture was stirred for 5 minutes, then allowed to warm to room temperature. The product was collected by filtration, washed successively with 50 ml. portions of 10% hydrochloric acid, acetone and water, and dried. A yield of 1.31 g. of pale tan solid was obtained. This had a melting point of 185° to 198°C. and analyzed 14.84% C, 81.20% S, and 1.25% H. Self-supporting films can be formed by heating the polymer at 190°C. and 10,000 lbs./in.^{2} pressure.

USE OF HYDROGEN SULFIDE AND SULFUR DIOXIDE

A process developed by R.J. Eckert and L.C. Fetterly (U.S. Patent 3,317,488; May 2, 1967; assigned to Shell Oil Company) results in sulfur-containing polymers prepared by reacting an ethylenically unsaturated compound simultaneously with hydrogen sulfide and sulfur dioxide in the mol ratio of $H_2S:SO_2$ of about 2:1.

The process may be performed throughout a relatively wide temperature range, and generally a temperature range wherein the H_2S and SO_2 react and combine to form sulfur is employed. Such a temperature range is normally from about -90°C. to about 50°C. with the preferred range being from about -10°C. to about 15°C. This preferred range could be effectively achieved by cooling the reaction with ice water. The process for preparing the new and valuable sulfur-containing polymers and their properties are illustrated by the following examples.

Unsaturated Monomer	Monomer Amount, ml.	Solvent	Reaction Phase	Yields, g.		Polymer, mol. wt.	Polymer Analysis			Remarks
				S	Polymer		C	H	S	
Cyclohexene	250	None	Liquid	---	30	---	11	2.0	87.1	
Styrene	100	200 ml. THF.	do	0	110	[2]3,800	45.1	3.8	49.9	Product soluble in benzene, odorless and colorless; very little free sulfur; resin soft point, 80–100° C.; resin decomposes at 250° C.
α-Methyl styrene	100	200 ml. THF.	do	40	60	[2]3,000	48.4	4.5	45.7	Resin softens at 80–100° C. and decomposes at 250° C.
Divinyl benzene	150	150 ml. THF.	do	Very little	---	---	51.9	4.8	41.0	Resin gelled, insoluble in solvents, decomposes at 170° C.
Ethyl acrylate	150	150 ml. THF.	do	Little	46	[3]3,000	35.1	4.2	46.3	Benzene-soluble, odorless, transparent yellow resin.
Butadiene	[1]3:2:1	None	Vapor	Some	---	---	13.5	2.4	72.5	Resin softening point: 65–70° C.; decomposes, 105–125° C.; insoluble in most inert solvents.
Do	[1]3:2:1	250 ml. THF.	Vapor-liquid.	None	60	---	20.2	4.1	62.0	Viscous brown, oil; sets to rubber upon standing.
Do	[1]1:4:2	250 ml. benzene.	do	Some	---	---	7.1	1.1	87.5	Yellow resin turns brown on standing.
Isoprene	100	200 ml. THF.	Liquid	None	40	---	32.1	4.3	60.7	Orange viscous oil which will cure rubber cements; sets to rubber solid in five days.
Cinnamyl alcohol	100	150 ml. THF.	do	45	135	[4]350	64.3	6.0	20.2	Yellow, viscous oil product.

[1] Molar ratio of butadiene: $H_2S=SO_2$.
[2] VP osmometer, dichloromethane.
[3] Ebulliscopically in benzene.
[4] VP osmometer, methyl ethyl ketone.

In general, the following procedure unless otherwise noted was employed throughout: A 500 ml., three-necked flask fitted with a high-speed stirring means, two gas inlet tubes and one gas outlet tube was utilized as the reactor. The rate of gas addition was determined by two flow meters. During the reaction the reaction flask was cooled with an ice-water mixture. The usual method was employed to isolate the reaction products, i.e., the reaction products were filtered and the filtrate evaporated to obtain the soluble reaction polymers.

Example: Into the above-described reactor were placed 300 ml. of allyl alcohol. Then under ice-water cooling and vigorous stirring, 2 molar parts of H_2S and 1 molar part of SO_2 were introduced simultaneously for one hour. The yellow precipitate (sulfur) was separated by filtration and dried (25 g.). The filtrate was evaporated in vacuo at room temperature and yielded 27 grams of a yellow resin. The resin became rubbery after a few days. If the introduction of the hydrogen sulfide and sulfur dioxide is continued for a longer time, the allyl alcohol-sulfur polymer will also separate together with the free sulfur. In such a case the mixture was dissolved in tetrahydrofuran (THF) using a Soxhlet unit. Then from this solvent, the free sulfur was separated by fractional crystallization. A small amount was purified by chromatography on aluminum oxide. (WOEIM, neutral, activity I) using THF as solvent. After evaporation of the solvent, the sample was dried over P_2O_5 and paraffin in vacuo at room temperature. The polymer had a softening point of 90° to 100°C. and decomposed at 170°C. The analysis of the polymer was: C, 27%; H, 4.4%; S, 57.5%.

REACTION OF SULFUR WITH LIQUID POLYTHIOPOLYMERCAPTAN POLYMERS

A process developed by E.R. Bertozzi (U.S. Patent 3,331,818; July 18, 1967; assigned to Thiokol Chemical Corporation) gives liquid polysulfide polymers, which, when cured to solid elastomers, are more resistant to solvents than are the cured conventional liquid polysulfide polymers. It also gives a curable, liquid polysulfide polymer which can be readily cured at room temperature upon exposure to the atmosphere without the use of a catalyst.

Liquid polysulfide polymers which can be cured at room temperature without the use of a catalyst and which have substantially greater solvent resistance in cured form than the cured conventional liquid polysulfide polymers may be formed by reacting the conventional liquid polymers with elemental sulfur to form a liquid polymer having —SSH terminals, a sulfur rank of about 1.6 to 5.0 and preferably about 2.5 to 3.5 and at least some sulfur linkages between the recurring hydrocarbon, oxahydrocarbon or thiahydrocarbon radicals which contain more than two sulfur atoms.

Conventional liquid polymers are those polythiopolymercaptan liquid polymers produced as described in U.S. Patent 2,466,963. These polymers have a molecular weight of about 500 to 12,000 and are liquid, i.e., pourable, at room temperature (about 25°C.). Structurally, they may be described as $HS(-R-S_x)_n-R-SH$ where the average of all x's is 1.5 to 2.0, n may be about 2 to 70 and R is a hydrocarbon, oxahydrocarbon or thiahydrocarbon radical such as $-C_2H_4-O-CH_2-O-C_2H_4-$. When these polymers are reacted with elemental sulfur, the reaction proceeds as follows:

$$HS(RS_r)_nRSH + (np + 2)S \longrightarrow HSS(RS_{r+p})_nRSSH$$

where the average of all r's is 1.5 to 2.0, p is 0.1 to 3.0, n is 2 to 70 and r+p is the desired sulfur rank; x, which, for any one linkage, may be 1 to 5 and for the average of all the linkages is about 1.6 to 5.0.

When packaged in an inert atmosphere such as under nitrogen the high sulfur rank liquid polymers are stable for prolonged periods of time. Upon exposure to the atmosphere, however, H_2S volatilizes from the polymer and the liquid polymer cures to a solid, solvent resistant elastomer. It is believed that the $-R-S-S-H$ terminated liquid polymers cure as follows:

$$2(R-S-SH) \longrightarrow (R-S_x-R) + H_2S$$

No curing agent is needed to facilitate the cures of these polymers at room temperature.

The uncatalyzed cures, however, require extended cure times. The polysulfide polymers can be cured more readily, however, with the aid of inorganic and organic peroxide curing agents such as lead peroxide which peroxides will also cure the conventional low sulfur rank polymers.

The high sulfur rank liquid polymers can be used to form solvent resistant films and coatings for roof coatings and other applications by exposing a film of the polymer to the atmosphere, in the presence or absence of one or more of the catalysts mentioned above, and allowing the polymer film to cure. The application of heat will expedite the cure. The cured polymers of the process are exceptionally resistant to solvents and exhibit a much greater resistance to solvents such as toluene than is shown by the cured conventional liquid polysulfide polymers.

The high sulfur rank liquid polymers can also be used to prepare curable sealants or caulking compositions in combination with the curing agents and the fillers, plasticizers, pigments and other adjuvants known to the art.

Examples 1 to 18:   In the following examples, the materials, in the amount indicated in the table, were charged into a three-necked, glass reaction flask. For the 200 gram polysulfide polymer charge, a 500 ml. flask was used; where 400 grams polysulfide polymer charge was employed, a one liter flask was used. The order of charge was liquid polysulfide polymer, sulfur, amine catalyst and water. The reaction was carried out under an atmosphere of nitrogen with continuous agitation. No external heat was applied and the reaction proceeded at the reaction temperature indicated in the example table. The reaction temperature was partially due to the heat of friction produced when the viscous reactants were agitated and partially to the exotherm produced by the reaction. The reaction was continued until the sulfur dissolved, the time varying from 2 1/2 to 11 hours as indicated. In Examples 1 to 14 the liquid polysulfide polymer used had essentially the following structure:

$$HS(C_2H_4-O-CH_2-O-C_2H_4-S-S)_{23}-C_2H_4-O-CH_2-O-C_2H_4-SH$$

while in Examples 15–18 the liquid polysulfide polymer had essentially the following structure:

$$HS(C_2H_4-O-CH_2-O-C_2H_4-S-S)_6-C_2H_4-O-CH_2-O-C_2H_4-SH$$

The reactions described above resulted in viscous materials which were usable as produced, the recovery being 100% in each case.

A portion of each of the prepared samples was charged into collapsible metal tubes to observe the stability of the polymer when protected from the atmosphere. When packaged in these sealed containers and protected from the atmosphere the uncured, high sulfur rank polymers remain stable for extended periods of time. Another portion of each of the prepared materials was formed into a thin bead to observe cure results when the material was exposed to the atmosphere. The results of these tests are indicated in the table.

The resultant samples can also be cured by using conventional liquid polysulfide polymer curing techniques as well as by the above mentioned method of exposing the polymer to the atmosphere. The following table, Examples 1 to 18, gives the results of the tests described previously.

Example No.	Liquid Polymer, gms.	Sulfur, gms.	Catalyst [1]		Water, cc.	Reaction Time, hrs.	Reaction Temp., °C.	Polymer, Sulfur Rank	Results Beads	
			Amt., cc.	Type					Skinned, hours	Cured, days
1	200	100	2	DBA	½	6½	22–35	4.60	1–2	14
2	200	50	2	DBA	½	3	23–53	3.30	1–2	14
3	200	100	10	DBA	½	6½	27–55	4.60	2	14
4	200	50	10	DBA	½	3	24–52	3.30	2	14
5	200	100	¼	DBA	½	6½	24–56	4.60	2	14
6	200	50	¼	DBA	½	5	40–50	3.30	2	14
7	400	20	4	DBA	1	2½	28–50	2.26	16	14
8	400	40	4	DBA	1	2½	22–46	2.52	16	14
9	400	20	20	DBA	1	3½	28–42	2.26	1	14
10	400	40	20	DBA	1	3½	28–51	2.52	1	14
11	400	20	4	TEA	1	2½	25–42	2.26	24	14
12	400	20	20	TEA	1	2½	25–42	2.26	24	14
13	400	40	4	TEA	1	2½	24–44	2.52	24	14
14	400	40	20	TEA	1	2½	24–40	2.52	24	14
15	200	50	2	DBA	½	8	24–37	3.30	24	14
16	200	100	2	DBA	½	10	28–42	4.60	24	12
17	200	50	¼	DBA	½	4	24–39	3.30	24	8
18	200	100	¼	DBA	½	11	24–35	4.60	24	7

[1] Abbreviations.—DBA=n-Dibutylamine. TEA=Triethylamine.

DISULFIDES FROM TETRATHIOCARBONATES AND ALKYL HALIDES

A process developed by R.W. Liggett (U.S. Patents 3,367,975; February 6, 1968 and 3,400,104; September 3, 1968; both assigned to Allied Chemical Corporation) results in organic disulfides by the reaction of an alkali metal tetrathiocarbonate with an organic halide. The disulfide reaction products are monomeric or polymeric compounds depending on whether the organic halide used is monofunctional or polyfunctional.

In general, an organic halide is mixed with a stoichiometric proportion of an alkali metal tetrathiocarbonate in a lower aliphatic alcohol reaction medium, and the mixture is heated at refluxing temperature for a period sufficient to bring about the desired reaction as illustrated below, usually a period between about 30 minutes and about 48 hours.

$$MS-\overset{\overset{\textstyle S}{\|}}{C}-S-SM + 2RX_n \longrightarrow R-S-S-R + 2MX_n + CS_2$$

where M in an alkali metal, R is an organic radical, X is a halogen selected from the group consisting of chlorine and bromine, and n is an integer from 1 to 3 inclusive.

Any organic halide that will react with the alkali metal tetrathiocarbonate at the reflux temperature of the alcohol solution may be employed provided that the halide is not so volatile that it may escape from the reaction mixture as a gas before reacting, and provided that it does not undergo simultaneous secondary or side reactions. The best seem to be mono-, di-, and trifunctional primary and secondary aliphatic chlorides and bromides of 1 to 20 carbon atoms containing in addition to the chlorine substituents only aliphatic hydrocarbon, or ether, or acetal groups.

Especially useful monofunctional halides are the primary aliphatic bromides and chlorides of the general formula

$$R^1CH_2X$$

where R^1 is alkyl of 1 to 20 carbon atoms and X is chlorine or bromine. Monofunctional secondary aliphatic bromides and chlorides containing 3 to 20 carbon atoms inclusive are also useful. Suitable di- and trifunctional halides include the 1 to 20 carbon atom aliphatic dibromides and dichlorides.

When the organic halide is monofunctional, the resulting reaction product is a simple monomeric compound. Thus such halides as methyl chloride or bromide, ethyl chloride or bromide, propyl chloride or bromide, butyl chloride or bromide, etc., result in the production of the monomeric disulfides:

$$CH_3-S-S-CH_3$$

$$CH_3CH_2-S-S-CH_2CH_3$$

$$CH_3CH_2CH_2-S-S-CH_2CH_2CH_3$$

$$CH_3CH_2CH_2CH_2-S-S-CH_2CH_2CH_2CH_3$$

and so on. When a di- or trifunctional organic halide is employed, i.e., when n is 2 or 3 in the above equation, the resulting reaction products are polymeric compounds and form repeating units of the character:

Polymer from difunctional halide — Polymer from trifunctional halide

The reaction proceeds as indicated below:

$$Na-S-\overset{\overset{S}{\|}}{C}-S-S-Na + X C_2H_4O CH_2O C_2H_4X \longrightarrow$$

sodium tetra-thiocarbonate — dihaloethyl formal

$$\left[-C_2H_4O CH_2O C_2H_4-S-S-\right] + 2NaX + CS_2$$

disulfide polymer

where X represents chlorine or bromine.

Example 1: The preparation of polymers is as follows. Six polymer preparations were carried out, in the first of which 237 grams (0.127 mol) of sodium tetrathiocarbonate were mixed with 22.0 grams (0.127 mol) of bis(2-chloroethyl) formal and 200 ml. of absolute ethyl alcohol. In the remaining preparations, various proportions of the dichloroethyl formal were replaced by trichloropropane. The mixtures were heated and stirred at reflux of 78 °C. under nitrogen for 24 hours, then allowed to stand at 25 °C. for 16 hours.

In each case, the product was a yellow liquid phase and a solid phase consisting of a yellow gummy solid mixed with a powdery yellow solid. The liquid phase was decanted from the solid, and the solid was washed with four 50 ml. portions of alcohol then with four 50 ml. portions of acetone. The solids were dried and weighed and then washed with water to dissolve the salt formed as a by-product in the reaction, and the insoluble residue was again dried and weighed. The alcohol solutions and the alcohol washings were combined and evaporated, and the residues weighed. The acetone-soluble fractions were similarly recovered and weighed after evaporation of the acetone.

In each case, the weight of the water-soluble fraction was in the range of 95 to 105% of the theoretical weight of the sodium chloride produced in the reaction. The theoretical yields and characteristics of the various polymer fractions are listed in Table 1.

TABLE 1: POLYMERS PREPARED WITH DICHLOROETHYL FORMAL AND SODIUM TETRATHIOCARBONATE

Polymer number	Percentage of formal replaced by trichloropropane	Molal ratio [1]	Yield of polymer fractions, percent [2]			Total yield of polymer, percent	Viscosity of polymer fractions		
			Alcohol soluble	Acetone soluble	Insoluble		Alcohol soluble	Acetone soluble	Insoluble
IV [3]	0	1.0:1.0:00	25	15	60	100	Thin liquid	Thin liquid	Plastic solid.
V	7.5	1.0:0.85:0.10	15	15	65	95	....do	....do	Elastic solid.
VI-1	0.3	1.0:0.997:.002	20	50	10	80	....do	Viscous oil	Sticky, plastic solid.
VI-2	0.3		20	20	55	95	Viscous oil	....do	Elastic solid.
VII-1	1.5	1.0:.985:.01	15	40	30	85	....do	....do	Plastic solid.
VII-2	1.5		25	20	55	100	....do	Very viscous oil	Sticky, plastic solid.

[1] Sodium tetrathiocarbonate:dichloroethyl formal:trichloropropane.
[2] Polymer yield calculated as percent of theory for a disulfide polymer.
[3] The designation is used for reference to the polymer fraction that is insoluble in alcohol and acetone.

Polymer IV, the linear polymer prepared by the reaction of sodium tetrathiocarbonate with dichloroethyl formal, was a light green solid with the consistency of putty. It gradually softened on heating and became quite fluid at 90° to 100°C. It was soluble in aniline and in dimethyl formamide. Its sulfur content was found to be 35.4%.

The infrared absorption spectrum of Polymer IV has characteristic absorption peaks at 3.4 to 3.5 microns, 6.8 microns, 7.1 microns, 7.3 microns, 7.9 microns, 8.3 microns, 8.7 microns, 8.8 microns, 9.1 microns, 9.3 microns, 10.2 microns, and 12.1 microns.

A polymer prepared in the same manner as Polymer IV above except that dibromoethyl formal was substituted for dichloroethyl formal had similar characteristics and a substantially identical infrared spectrum. Polymer V, the cross-linked polymer prepared by the reaction of sodium tetrathiocarbonate with dichloroethyl formal and trichloropropane (1.00:0.925:0.05 mol ratio), was a gray rubbery solid, which softened and lost its elasticity at 210° to 225°C., melted at 240°C. and decomposed at 260° to 280°C. It was swollen by aniline, but not dissolved. It was soluble in dimethyl formamide. Its sulfur content was 32.1%.

Example 2:   The liquefaction of polymers is as follows. Polymers IV and V from Example 1 above were mixed with hot (80°C.) alcoholic sodium hydrosulfide (NaHS) in the proportion of about one part of polymer per ten parts of 5% alcoholic solution of hydrosulfide. Upon agitation of the mixtures for about 30 minutes, Polymer IV became a free flowing liquid, Polymer V became a plastic mass in which the original particles had lost their identity.

A sample of Polymer IV was liquefied by heating a suspension of a one gram sample in 10 ml. of an aqueous solution containing about 5% sodium hydrosulfide and about 2% sodium sulfite resulting in a fluid polymer. Treatment of samples of Polymers IV and V with 10% aqueous sodium hydroxide at 100°C. for 2 hours did not change the viscosities of the polymers, and treatment of these polymers with aniline under similar conditions had only slight if any effect on their viscosities.

The infrared absorption curve for the liquid polymer obtained by treating Polymer IV with sodium hydrosulfide exhibits characteristic absorption peaks at 3.4 microns, 3.5 microns, 6.8 microns, 7.1 microns, 7.3 microns, 7.7 microns, 8.7 microns, 9.0 microns, 9.1 microns, 9.3 microns, and 10.2 microns.

Example 3:   The curing of liquefied polymers is as follows. One hundred parts each of the liquid polymers obtained from Polymer IV above, were mixed with 30 parts each of a paste prepared from 50 parts of lead dioxide and 45 parts of dibutyl phthalate plasticizer and 5 parts of stearic acid, and a similar mixture was made from a commercial polysulfide polymer prepared by the reaction between polysulfide of rank 2.25 and bis(2-chloroethyl) formal containing 1.5% trichloropropane, followed by liquefaction with sodium hydrosulfide and sodium sulfite. The mixtures were air cured for 16 hours and resulted in the formation of rubbery products. When tested for recovery from compression by a qualitative test, it was observed that the cured polymers of the example were superior to the commercial product.

Example 4:   The preparation of disulfide monomer is as follows. In 80 ml. of ethanol there were mixed 13.7 grams (0.10 mol) of n-butyl bromide and 9.3 grams (0.05 mol) of sodium tetrathiocarbonate and the resulting mixture was refluxed for 2 hours at 78°C.

The alcohol was evaporated from the mixture in a rotary evaporator, and there was obtained a mixture of a salt and an oily liquid. The oily liquid was extracted from the salt with petroleum ether, and recovered from the extract by evaporation of the petroleum ether. The yellow oil remaining after evaporation of the solvent weighted 7.1 grams, had a refractive index of 1.4915, and it boiled at 94° to 95°C. at 4 mm. Hg. Elemental analysis of a distilled sample is shown below and compared to calculated values for n-butyl tetrathiocarbonate and for n-butyl disulfide.

	Found	Calculated	
		n-Butyl tetrathio-carbonate	n-Butyl disulfide
Percent:			
S	37.2	50.4	36.0
H	10.2	7.1	10.1
C	53.1	42.5	53.0

These results indicate that the compound obtained was n-butyl disulfide.

$$CH_3CH_2CH_2CH_2-S-S-CH_2CH_2CH_2CH_3$$

obtained in a yield of 80% of theory.

The polymer made in the process is characterized by the presence of the following repeating units:

$$+C_2H_4OCH_2OC_2H_4-S-S+$$

BLOCKED LIQUID POLYSULFIDE POLYMERS

In U.S. Patent 3,331,818 there is disclosed a high sulfur rank, —SSH terminated polysulfide polymer which cures at room temperature without the use of a cure catalyst upon exposure to the atmosphere. These cured polymers are more resistant to solvents such as toluene than the cured conventional liquid polymers. These high sulfur rank polysulfide polymers have a disadvantage, however, in that during the cure thereof they emit noxious hydrogen sulfide fumes. For some applications of these polymers, such as sealant or coating usage in con- confined quarters, the noxious fumes emitted present a health hazard when large quantities of the polymer are used.

According to another process developed by E.R. Bertozzi (U.S. Patent 3,422,077; January 14, 1969; assigned to Thiokol Chemical Corporation), the emission of noxious fumes during the cure of high sulfur rank, —SSH terminated polysulfide polymers can be avoided by blocking the —SSH terminals, prior to the cure of the polymers, with an aldehyde or a ketone. The blocking reaction can be conducted either before, after or concurrent with the formation of the high sulfur rank polymers from the conventional liquid polysulfide polymers of U.S. Patent 2,466,963.

The blocking of the —SSH terminals is accomplished by reacting the high sulfur rank, —SSH terminated polysulfide polymer with an aldehyde or ketone in the presence of an amine. It is believed that the aldehyde or ketone reacts with the amine to form an amine hemi-acetal or hemi-ketal terminal with the —SSH group as indicated by the following re- action with an aldehyde:

$$-R-SSH + R'-CHO \xrightarrow{\text{amine}} -R-SS-\underset{\underset{R'}{|}}{CH}-OH$$

The blocking of the terminal groups can be accomplished with the aldehyde or ketone blocking agents either before, after or during the formation of the liquid, high sulfur rank polysulfide polymers from the conventional, liquid polysulfide polymers. The high sulfur rank polymers can thus be blocked after they are formed by reacting the blocking agent in the presence of an amine or they can be blocked concurrently with their formation by adding the blocking agent to a sulfur/amine/conventional liquid polysulfide polymer reaction system. The blocked, high sulfur rank polymers can also be formed by first block-ing a conventional, liquid polysulfide polymer with the aldehyde or ketone blocking agent in the presence of an amine and subsequently reacting the blocked conventional liquid polymer with sulfur to form a blocked, high sulfur rank liquid polymer. Slightly more of the amine is usually needed where sulfur is added to a blocked conventional liquid polymer.

Where sulfur is added to the blocked, conventional liquid polymer, little or no H_2S evolves. Where sulfur is added to the unblocked polymer, or where sulfur, conventional liquid poly-mer and blocking agent are reacted simultaneously noticeable amounts of H_2S evolve. It also usually requires longer to add the sulfur to the blocked conventional polymer than it does to the unblocked conventional polymer. No matter how the blocked high sulfur rank poly-mers are prepared, no external heating is required to promote any of the reactions.

The temperature of the reaction systems, however, will climb from room temperature to about 40° to 60°C. due to frictional heat caused by the agitation or stirring of the viscous reaction systems and/or by the exotherm produced by the reactions. The reactions are pref-erably carried out under an inert atmosphere, such as under a blanket of nitrogen. The aldehydes which may be used in the blocking reactions include formaldehyde, furfural, and acetaldehyde and they should be present in about 2 to 9% by weight of liquid polymer. The preferred of these aldehydes is formaldehyde.

The ketones which may be used in this blocking reaction include acetone and they should be present in approximately 10% by weight of liquid polymer. Amine catalysts which can be utilized in the blocking process of this method include triethylamine, dibutyl amine and n-butyl amine. The preferred of these catalysts is triethylamine and they should be present in 0.5 to 10% by weight and preferably about 0.5% by weight of liquid polymer.

With several possible variations relating to the carbonyl compound used and operating con-ditions, the process follows the procedure given below.

Example: A three-necked 2.5 liter glass reaction flask was charged with 1,000 grams of liquid polysulfide polymer having essentially the formula

$$HS(C_2H_4-O-CH_2-O-C_2H_4-S)_{23}-C_2H_4-O-CH_2-O-C_2H_4-SH$$

289.5 grams of powdered sulfur, N.F., 45 grams of p-formaldehyde, 1.25 ml. of triethyl-amine and 2.5 ml. of distilled water. The mixture was reacted with continuous agitation under a nitrogen atmosphere for 10 hours. No external heat was applied; however, the heat produced by the agitation and reaction exotherm caused the temperature to rise to 40° to 50°C. At the completion of the reaction time, as evidenced by a clearing of the color of the reaction mixture, the product was a blocked, liquid polysulfide polymer having a sulfur rank of 3.5. The viscosity of this liquid polymer was 340 poises. A portion of this high rank polymer was cured overnight at room temperature using the following formulation:

	Parts by Weight
Liquid polymer	100
Essex SRF #3 (semi-reinforcing furnace black)	30

(continued)

	Parts by Weight
Durez 10694 (phenolic adhesion additive)	5
Curing agent	15

The polymer cured without the emission of H_2S. The curing agent had the following formulation:

	Parts by Weight
Lead dioxide	50
Stearic acid	5
Dibutyl phthalate	45

The cured polymer had the following physical properties:

Hardness—durometer Shore A	
After 3 days	47
After 6 days	53
Tensile, psi	297
Elongation, percent	470
Modulus, percent:	
200	165
300	232

The solvent swell of the polymer prepared and cured as described above, after 24 hours in toluene, was 94% as compared to a solvent swell in toluene of 136% for a conventional liquid polysulfide polymer having a sulfur rank of 2.0 and cured as indicated above.

MODIFIED POLYSULFIDE POLYMERS

POLYSULFIDES WITH ETHER GROUPS

Three processes in the early and mid 1950's deal with polysulfides containing ether groups. All were developed by F.K. Signaigo (U.S. Patent 2,586,182; February 19, 1952; U.S. Patent 2,685,574; August 3, 1954; and U.S. Patent 2,700,685; January 25, 1955; the latter two patents both assigned to the U.S. Secretary of the Navy).

A mixture of 252 parts (1 mol + 5%) of sodium sulfide ($Na_2S \cdot 9H_2O$), 68 parts (2 mols + 5%) of sulfur and 90 parts of water is stirred at 50°C. until all the sulfur is dissolved. To form magnesium hydroxide, which acts as a dispersing agent in the subsequent polymer formation, 12 parts of sodium hydroxide are added, followed by the gradual addition with stirring of a solution of 30.6 parts of magnesium chloride ($MgCl_2 \cdot 6H_2O$) in 40 parts of water. To the solution of sodium polysulfide maintained at about 50°C. is then added with stirring, over a period of 3 to 4 hours, 246 parts (1 mol) of 2,3-dibromopropyl ethyl ether (prepared by brominating allyl ethyl ether in chloroform).

The polymeric sulfide which separates is digested by stirring the reaction mixture at 70° to 75°C. for 7 to 8 hours. After cooling, the contents of the reaction vessel are poured into a large volume of water, and the solid is washed with water by decantation until the supernatant layer becomes pale yellow, then air dried.

The processes produce, in essentially the same manner by varying the ether compound, a (methoxymethyl) ethylene polysulfide, the ethoxyethyl polymer, and the n-butoxymethyl polymer.

POLYSULFIDES MODIFIED WITH ARSENIC

A process developed by L. Akobjanoff (U.S. Patent 2,587,805; March 4, 1952) gives a modified polysulfide plastic containing the unit:

where

represents a structure selected from the groups consisting of two carbon atoms connected by a valence bond and two carbon atoms separated by and joined to the intervening structure, and where:

$$\overset{\displaystyle \overset{(Z)_a}{\|}}{-Z-B-(Z+_b}$$

represents the residue of the chalcogen-containing compound, in which Z is a chalcogen and B is a nonreactive nonsulfur-containing nucleus. The values of a and b depend on the valence of B. The relationship between the values of a and b and the valence of B is shown as follows:

Valence of B:	2	3	4	5	6
Value of a:	0	0	1	1	2
Value of b:	1	2	1	2	1

Sulfur groups such as:

$$-S-S-; \quad \overset{\displaystyle \overset{S}{\|}}{-S-S-}; \quad \overset{\displaystyle \overset{S}{\|}}{\underset{\displaystyle \underset{S}{\|}}{-S-S-}}; \quad \overset{\displaystyle \overset{S}{\|}}{\underset{\displaystyle \underset{S}{\|}\underset{S}{\|}}{-S-S-}}; \quad or \quad \overset{\displaystyle \overset{S}{\|}\overset{S}{\|}}{\underset{\displaystyle \underset{S}{\|}\underset{S}{\|}}{-S-S-}}$$

are known to exist in the repeating units of polysulfide plastics and are the cause of the disagreeable odors possessed by these plastics. The presence of a nonreactive nonsulfur nucleus in these sulfur groups stabilizes the polysulfides against loss of volatile fragments containing sulfur, as shown by the odorless property of these new plastics. Also these plastics do not have a tendency to blister or become porous when being processed as by molding.

The nonreactive nucleus is made from the elements of the group consisting of As, C, Hg, Mo, Sb, Se, Sn, Te, V, W and Zn. The actual atomic grouping of B will depend upon the valence of the element being used and the conditions of reaction employed in producing the chalcogen-containing compound. For example, when sodium sulfide is reacted with selenium a compound having the following structural formula is produced: Na—S—Se—Se—Se—Na which corresponds to the general formula MZ—B—Z'M in which B is —Se—Se—, M is sodium, Z is sulfur and Z' is selenium. To produce a chalcogen compound corresponding to the formula:

$$MZ'-B\overset{\displaystyle \diagup Z'M}{\diagdown Z''M}$$

sodium sulfide and arsenious sulfide (As₂S₃) are reacted together to produce the compound:

$$NaS-As\overset{\displaystyle \diagup SNa}{\diagdown SNa}$$

A compound corresponding to the formula:

$$MZ'-\underset{\underset{Z''''}{\overset{\overset{Z''}{\|}}{B}}-Z'''M$$

is prepared by reacting sodium sulfide with tin disulfide (SnS_2) to yield the substance:

$$NaS-\underset{\overset{\overset{S}{\|}}{Sn}}-SNa$$

A compound corresponding to the formula:

$$MZ'-\underset{\overset{\overset{Z''}{\|}}{B}}{\overset{Z'''M}{\diagdown}}{Z''''M}$$

is prepared by reacting sodium sulfide with arsenious pentasulfide (As_2S_3) to give:

$$NaS-\underset{\overset{\overset{S}{\|}}{As}}{\overset{SNa}{\diagdown}}{SNa}$$

A compound corresponding to the formula:

$$MZ'-\underset{\underset{Z''''}{\overset{\overset{Z''}{\|}}{B}}{\|}}-Z'''M$$

is prepared by reacting sodium sulfide with ammonium molybdate $(NH_4)_6Mo_7O_{24}$ to produce the compound:

$$NaS-\underset{\underset{S}{\overset{\overset{S}{\|}}{Mo}}{\|}}-SNa$$

Thus, in the production of these plastics, the chalcogen atoms are chemically bound together by the nonreactive nucleus B.

A number of polymers containing nuclei such as —Se—Se—, —Te—Te—, —Hg—, and many others can be made by this process. Three arsenic-containing polymers can be made as follows.

Example 1: A compound containing the nucleus —As< was prepared by melting a mixture comprising 7.8 parts of Na_2S with 25 parts of As_2S_3 prepared by reacting sodium arsenite with H_2S in acid solution, and extracting the resulting mixture with acetone to eliminate free sulfur. The resulting compound was sodium sulfoarsenite having the formula $(NaS)_3As$. A modified plastic having the general formula is shown on the following page.

$$\left[\begin{array}{c} -S \\ \diagdown \\ AsSR- \\ \diagup \\ -S \end{array} \right]_y$$

It was prepared by reacting 48 parts of $(NaS)_3As$ with 30 parts of ethylene dichloride and 180 parts of water at 120°C. for 15 hours. The resulting product was granular and white in color, workable into a plastic sheet on a rubber mill; possessed no odor even when heated at a temperature of 150°C. and above; slightly thermoplastic at 185° to 195°C.; infusible even at 250°C.; soluble in carbon disulfide and insoluble in benzene.

A further reaction was carried out in which 72 parts of $CH_3CH(OC_2H_4Cl)_2$ was reacted with 48 parts of $(NaS)_3As$ in the presence of 180 parts of water at 100°C. for 16 hours. The resulting product was a translucent greenish, tacky, very elastic mass, having no odor, soluble in carbon disulfide, benzene and acetone, and insoluble in naphtha.

A still further reaction was carried out in which 70 parts of C_2H_5Br was reacted with 48 parts of $(NaS)_3As$ in the presence of 180 parts of water at 35° to 40°C. for 20 minutes. The resulting product was a yellowish waxy substance soluble in water, alcohol and insoluble in ether, acetone and chloroform; it decomposes at 120° to 130°C.

<u>Example 2:</u> A mixture of compounds containing the nuclei:

$$-SAs\begin{array}{c} S- \\ \\ S- \end{array} \quad , \quad -SAs\begin{array}{c} S- \\ \\ O- \end{array} \quad , \quad \text{and} \quad -SAs\begin{array}{c} O- \\ \\ O- \end{array}$$

was prepared by reacting 14 parts of arsenic trioxide As_2O_3 with 15.8 parts of Na_2S in 180 parts of water. The resulting product contained sodium sulfoarsenite $NaSAs(SNa)_2$; sodium disulfomonoxyarsenite:

$$NaSAs\begin{array}{c} SNa \\ \\ ONa \end{array}$$

and sodium monosulfodioxyarsenite:

$$NaSAs\begin{array}{c} ONa \\ \\ ONa \end{array}$$

A modified plastic containing compounds having the general formula:

$$\left[As\begin{array}{c} SR- \\ \\ SR- \\ \\ OH \end{array} \right]_y \quad ; \quad \left[As\begin{array}{c} SR- \\ \\ OH- \\ \\ OH \end{array} \right]_y \quad \text{and} \quad \left[As\begin{array}{c} SR- \\ \\ SR- \\ \\ SR- \end{array} \right]_y$$

was produced by reacting 63.2 parts of the above mixture of sulfo- and sulfooxyarsenites with 60 parts of ethylene dichloride in emulsion in 494 parts of water, 6 parts of a 20% solution of ammonia, 6 parts of sodium sulforicinate and 3 parts of casein for 12 hours at 115°C. A granular condensate was obtained.

Example 3: A compound containing the nucleus —As(S)< was prepared by melting a mixture of 7.8 parts of Na_2S with 32 parts of As_2S_5 obtained by reacting Na_3AsO_4 in acid water solution with H_2S at a low temperature. The resulting compound was sodium sulfoarsenate Na_3AsS_4. A modified plastic having the general formula:

$$\left[\begin{array}{c} -S \\ \diagdown \\ \qquad As(S)SR- \\ \diagup \\ -S \end{array} \right]_y$$

was prepared by reacting 54.4 parts of Na_3AsS_4 with 30 parts of ethylene dichloride in the presence of 180 parts of water at 120°C. for 15 hours. The resulting product was an elastic composition workable on a rubber mill, substantially odorless, thermoplastic at 100°C., soluble in carbon disulfide, and insoluble in benzene.

METALLOTHIOPLASTS

Another process developed by L. Akobjanoff (U.S. Patent 2,614,097; October 14, 1952) gives polymers having the same basic structure as those described in U.S. Patent 2,587,805. In this process,

$$\begin{array}{c} (Z')_a \\ \ddot{} \\ -Z-\ddot{B}-(Z''-)_b \end{array}$$

represents the residue of the chalcogen-containing compound, in which Z, Z', and Z" are chalcogens and B is a nonreactive nucleus chosen among the metals: Hg, Mo, Sb, Sn, V, W, Zn, known to be the only ones able to form thio-(seleno-, telluro-) acid salts which dissolve in water without being hydrolized.

The process for making the metallothioplasts is essentially the same as the one described in U.S. Patent 2,587,805.

POLYTHIOUREA/POLYSULFIDE RESIN

A process developed by G.L. Wesp (U.S. Patent 2,957,845; October 25, 1960; assigned to Monsanto Chemical Company) gives resins by cocuring mixtures of linear polythiourea and polysulfide polymers. Depending on the particular curing agent or agents used and to a certain extent on the particular polymer blend being cured, curing can be carried out at room temperature or below but normally at elevated temperature. Usually temperatures as high or higher than 175°C. should not be used in curing since substantial decomposition of the polysulfide polymer component will normally result, and curing should usually be carried out at about 150°C. or lower.

To make the polythiourea, a tared liter flask fitted with a nitrogen inlet, Trubore stirrer, thermometer, dropping funnel and reflux condenser vented through a mineral oil trap, was purged with nitrogen and charged with 265.8 g. (1.508 mols) of 1,2-bis(3-aminopropoxy) ethane and 532 ml. of methanol. Exothermic heat of mixing was noted. The mixture was cooled in an ice bath, the nitrogen purging stopped, and with good agitation 120.6 g. (1.584 mols), 5% excess, of carbon disulfide was added gradually via the dropping funnel, as follows:

Minutes	°C.	CS_2, ml.	Remarks
0	2	0	Start.
10	16.5	34	
15	19	50	Clear, faint green.
20	18.7	64	Slightly milky.
30	17	95	All CS_2 in.
31	16.5	-	Suddenly, opaque white.

After cooling to 4°C. during 1 hour, the stirrer was stopped, and the mixture separated into layers. The upper layer (about 33% of the total) was clear, almost colorless and very fluid. The bottom layer was clear, very light yellow and viscous. On stirring and warming to 25°C., the mixture became homogeneous; on recooling to 5°C., phase separation reoccurred. The supernatant layer was decanted and the residue was stored overnight under nitrogen at room temperature. The flask was then fitted with a nitrogen bubbler, placed in an oil bath, and the condenser was changed from reflux to distillation position.

Solvent was then removed by distillation with the stirrer operating. Distillation was essentially complete after 35 minutes, the oil bath temperature going from 98° to 128°C. At this point slow evolution of hydrogen sulfide began and continued at an increasing rate during the next 50 minutes, as the bath temperature was increased to 170°C. After 5 more minutes at 170° to 174°C. gas evolution essentially stopped, and the viscous, clear, light green polymer was bubbled with nitrogen and stirred and heated during the next 3.5 hours, at a 180° to 195°C. bath temperature.

After cooling the resin under nitrogen to about 140°C., 0.75 g. of 2-mercaptobenzothiazole (an antioxidant) was stirred in, and the hot viscous mass was dumped into a glass tray, greased with silicone grease. The cooled product was a tacky, clear, very light amber semisolid. The specific viscosity of the product at 1% by weight concentration in dimethylformamide at 25°C. was 0.263.

The polysulfide polymer used in most of the experimental blends was "Thiokol" LP-2, marketed by the Thiokol Corporation. This polymer in undiluted form is a viscous amber liquid having a viscosity of about 400 poises at 25°C., a specific gravity of about 1.27, and a molecular weight of about 4,000.

A different type of polysulfide polymer of unknown structure was used in making up one resin. This resin is "Thiokol" FA. Typical properties of this polymer are that it has the appearance of a crude rubber which is light brown in color and has a specific gravity of 1.34. The table on the following page describes curing tests on a number of different blends of polythiourea and polysulfide polymer. In these tests the polythiourea polymer was blended with the polysulfide polymer by mixing these polymers at temperatures in the range of 50° to 80°C. After the polymer mixture had cooled to room temperature the curing agents were then mixed into the blended polymer.

Curing Tests

Blend No.	PTU, No.	PS, PHR	Additives, PHR				Curing Time At 25° C., hours
			PF	ZnO	PbO₂	SA	
1	3	100					No Cure (24 hrs.).
2	3	100	8				3.
3	3	100	7.5	20			1.
4	3	100			20		0.25.
5	3	100	10		20		0.17.
6	3	100	10		10		0.33.
7	3	100	4		10	2	0.42.
8	3	100	4		10		0.33.
9	3	100	4	20			16.
10	1	200	1		14	3	24.
11	1	200	2		14		0.30.
12	1	300	2		21		0.30.

PTU—Polythiourea polymer.
PS—Polysulfide polymer, which was "Thiokol" LP-2.
PHR—Parts per hundred parts based on the polythiourea.
PF—Paraformaldehyde.
SA—Stearic acid.

The PTU, polythiourea polymers used in the table are described in detail and the PS, polysulfide polymer used in this table was the "Thiokol" LP-2 polymer. All the additives described in the table above are curing agents with the exception of the stearic acid which is a curing inhibitor designed to slow down the curing process. The compounded mixture in each case containing the curing agents or agent was spread in a 1/16 inch layer on a chrome plate metal surface and allowed to cure at room temperature, i.e., about 25°C. The time to cure in each instance was reported as the time required for the polymer to take on rubbery characteristics.

POLYSULFIDES HAVING SPIROBI(META-DIOXANE) GROUPS

A process developed by H.A. Stansbury, Jr. and H.R. Guest (U.S. Patent 2,960,495; November 15, 1960; assigned to Union Carbide Corporation) gives polymers containing spirobi(meta-dioxane) and sulfur groups. It is also a method for producing polymers containing substituted spirobi(meta-dioxane) groups connected by sulfur linkages which are useful as intermediate reactants and as accelerators for curing rubber.

A solution of 3,9-divinylspirobi(meta-dioxane) (170 g.; 0.8 mol) in an equal weight of diethyl ether of diethylene glycol solvent was stirred at 0° to 10°C., while dry hydrogen chloride gas was fed until 59 g. (1.6 mols) was absorbed.

A mixture of sodium sulfide monohydrate (240 g.; 1 mol), sulfur (96 g., 3 atomic weights) and water (99 ml.) was stirred and heated to form 1 mol of sodium tetrasulfide in aqueous mixture. A solution of magnesium chloride catalyst (7 g.) in water (10 ml.) was added to the tetrasulfide mixture.

The resulting mixture was stirred and heated at a temperature of 80° to 85°C. while the diethyl ether of diethylene glycol solution was added dropwise over a period of one hour. After a reaction period of 2.5 hours at 100°C., the mixture was poured into water (2 liters) with stirring. Acetic acid (47 ml.) was added to make the mixture weakly acidic. The solid product was filtered off, washed with methanol and dried under vacuum at 30°C. The dry, rubber-like polymer was insoluble in hot acetone and in hot toluene. The overall yield was 90% based on 3,9-divinylspirobi(meta-dioxane). Analysis — Calc. for $C_{11}H_{18}O_4S_4$: C, 38.6; H, 5.3; S, 37.4. Found: C, 37.8; H, 6.0; S, 31.0.

Dioxane can be used as the solvent instead of diethyl ether of diethylene glycol. After the gummy polymeric product was washed with methanol it was dried at room temperature under vacuum. Analysis — Calc. for $C_{11}H_{18}O_4S_4$: C, 38.6; H, 5.3; S, 37.4. Found: C, 37.4; H, 6.5; S, 26.5.

This process illustrates the ability of polysulfided spirobi(meta-dioxane) polymers to be cross-linked. The polymer from above was cross-linked to a slight extent with zinc oxide and to a considerable degree with tert-butyl peroxide.

	A	B
Formulation grams:		
polymer described previously	30	30
zinc oxide	3	-
tert-butyl peroxide	-	0.3
stearic acid	0.15	0.15
Milling temperature, °C.	30	30-160[2]
Milling time, minutes	5	10
Curing temperature, °C.	140	140
Curing time, minutes	45	45
Physical properties on cured compositions:		
tensile, psi	too weak	700
elongation, percent	too weak	300
load at 100% elongation, psi	-	700
hardness, Durometer A	8[1]	85
hardness, Durometer D	-	31

[1] Same hardness as polymer described previously.
[2] Partially cross-linked on mill.

To show that polysulfided spirobi(meta-dioxane) polymers are effective accelerators for the curing of rubber, the following formulations were compounded on a two-roll mill at 40° to 75°C.

(a)   100 g. of natural rubber
1 g. of "Age Rite" powder antioxidant
5 g. of zinc oxide
2 g. of sulfur
3 g. of stearic acid
50 g. of "Kosmobile 77EPC" carbon black

(continued)

1 g. of polymer made with dioxane.
(b) Same as (a) except the polymer was omitted.

Both (a) and (b) were cured 60 minutes at 140°C. respectively in positive molds under pressure. The cured compositions had the following properties:

	(a)	(b)
Tensile, psi	700	380
Elongation, percent	400	375
Load at 300% elongation, psi	400	200
Hardness, Durometer A	35	14

ALKYLENE SULFIDE/PROPYLENE OXIDE COPOLYMERS

A process developed by A.E. Gurgiolo (U.S. Patent 3,000,865; September 19, 1961; assigned to The Dow Chemical Company) results in copolymers containing in combined form from 5 to 95 weight percent of propylene oxide and the remainder the alkylene sulfide. These copolymers are white to yellowish solid materials that have at least one and usually more of a variety of uses including the preparation of moldings, films, fibers and in coating applications.

They generally provide high strength fabricated articles that are possessed of good dielectric characteristics and a good stability to heat and light, especially the copolymers containing from 80 to 90 weight percent of propylene oxide and from 20 to 10 weight percent of the selected alkylene sulfides. They have an average molecular weight in excess of 100,000, a softening point generally above 60°C. and a melting point over 100°C. They are insoluble in and resistant to water and aqueous acids and alkalies, but are soluble in aromatic hydrocarbons and most oxygenated organic solvents.

The copolymers may be made by the copolymerization of propylene oxide and the selected alkylene sulfide in the presence of a ferric chloride-propylene oxide complex catalyst. This catalyst appears to be a mixture of complex salts containing ferric chloride and propylene oxide in definite molecular ratios. Details of the preparation and purification of the catalyst are given in U.S. Patent 2,706,181.

In the copolymerization of the propylene oxide and the selected alkylene sulfide, the monomers and the catalyst may be simply mixed together and charged into a closed vessel and heated until the polymerization is complete. It is usually beneficial for the reaction mass to be agitated during the polymerization. The amount of catalyst employed is about 4 weight percent.

The copolymerization may be carried out within the temperature range of 30° to 150°C. At the lower temperatures higher yield of the solid copolymers may be realized but the polymerization time is generally longer and may often be 200 hours or more. At the high temperatures, the rate of reaction is relatively rapid and a suitable point for the termination of the reaction may be reached in less than 3 hours. However, at these high temperatures, the yield of the solid resin obtained may decrease. To obtain a fairly rapid rate of reaction with a suitable yield of the desired solid polymer, the copolymerization is ordinarily carried out at a temperature between 60° and 100°C., the optimum temperature being about 80°C. At the temperatures of 60° to 100°C., the copolymerization usually is substantially completed in about 18 to 120 hours, the optimum being from 40 to 60 hours.

The copolymerization of propylene oxide and the selected alkylene sulfide may also be carried out in an inert nonaqueous diluent medium. The employment of such a medium for the polymerization may sometimes tend to reduce the rate of the reaction, although, in certain instances, it may facilitate the achievement of a more nearly complete copolymerization of the monomers. The medium either may be a solvent or a nonsolvent suspending medium, so long as it boils at about the desired polymerization temperature.

In this way, the utilization of reflux techniques permits an easy means for the regulation of the reaction temperature. While various low boiling liquid, nonsolvent media may also be employed, it is usually more desirable to utilize solvents. The inert nonaqueous diluent medium may generally be used in a quantity that is approximately equal to the quantity of the monomers being copolymerized.

Several procedures may be used for the recovery and purification of the copolymerized product from the reaction mass. For example, the unreacted monomers and the solvent or other diluent medium (when one has been employed) may be stripped from the reaction mass by vaporization to leave the catalyst-containing copolymeric material. The crude copolymer is in the form of a resilient solid mass having a brownish to blackish coloration and may be associated with liquid polymers that may have been formed during the reaction.

Usually the impure solid copolymer may be dissolved in a suitable solvent, such as hot acetone, which may then be acidified with a hydrohalic or other suitable acid to convert the iron-containing catalyst to a soluble salt form before precipitating the solid polymer by crystallization from the solution at a low temperature, generally about −20°C. or below. Recrystallization may be employed for further purification until a suitable solid copolymeric material is obtained that has a sufficiently high molecular weight to not soften excepting at temperatures that are in excess of about 60°C. Alternatively, water may be added to the acetone solution of the crude polymer from the reaction mass in order to precipitate the iron as a hydroxide which may be removed by filtration before precipitation of the purified copolymeric material.

POLYSULFIDE POLYMER CONTAINING STYRENE OXIDE

A process developed by L. Montesano (U.S. Patent 3,101,326; August 20, 1963; assigned to Bell Telephone Laboratories, Incorporated) gives compositions comprising an organic polysulfide and styrene oxide where the offensive odor of the polysulfide is effectively masked. There may also be included in these compositions a polyepoxide compound formed by the reaction of a dihydric phenol with a 1,2-epoxy compound; such compositions are of particular interest for use as a cable sealant. The polysulfides may be any of those marketed commercially under various trade names, one such group being marketed by the Thiokol Corporation and including "Thiokol LP-3," "Thiokol LP-8" and "Thiokol LP-33." These commercial resins may be represented by the general formula:

$$HS(CH_2)_2[(OCH_2)_2CH_2S_2\text{———}(CH_2)_2]_x(OCH_2)_2CH_2SH$$

where x is an integer such that the total molecular weight is approximately 1000. They differ primarily in molecular weight, cross-linking and viscosity as evidenced by the following data.

Resin	Molecular weight	Percent cross-linking	Viscosity in centipoises at 27° C.
Thiokol LP-3	1,000	2	700-1,200
Thiokol LP-8	500-700	2	250-350
Thiokol LP-33	1,000	0.5	1,300-1,500

An organic polysulfide suitable for use in forming gels, and, more particularly for use in cable-plugging mixtures may be prepared according to U.S. Patent 2,402,977.

A material even more suitable for the production of cable plugs is obtained by the inclusion of up to about 2 mol percent of trichloropropane with the dichlorodiethyl formal used in making the polymer. Cross-linked structures and structures capable of cross-linking are thereby obtained. These structures are too complex to permit a single representation by formulas but are characterized by additional SH groups in other than terminal positions.

To effectively mask the odor of the polysulfides styrene oxide is employed in an amount within the range of 5 to 30 parts by weight per 100 parts by weight of polysulfide. Although a value of 5 parts by weight is indicated as a lower limit, values less than 5 may be employed depending upon the particular polysulfide employed. The use of greater than 30 parts by weight per 100 parts by weight of polysulfide may be tolerated, however, such quantities fail to produce any further perceptible effect in masking the offensive odor.

The polyepoxy compounds useful for forming gels are well-known and described, for example, in U.S. Patent 2,506,486. They are either monomeric or partially polymerized forms of a diglycidyl ether of a diphenol, commonly prepared by reacting 2 or more molar proportions of epichlorhydrin with one molar proportion of a diphenol. The materials may be represented by the formula:

$$CH_2\text{———}CH\text{—}CH_2\text{—}(O\text{—}R\text{—}O\text{—}CH_2\text{—}\underset{\underset{OH}{|}}{CH}\text{—}CH_2)_n\text{—}O\text{—}R\text{—}O\text{—}CH_2\text{—}CH\text{———}CH_2$$

where R is an aromatic-bearing radical which may vary considerably in nature and n is an integer generally within the range of 5 to 9. Diphenols suitable for reacting with epichlorhydrin are, as described in U.S. Patent 2,506,486 of the general formula:

in which the phenolic hydroxy groups may be in the 2,2'; 2,3'; 2,4'; 3,3', 3,4' positions on the aromatic rings. The equivalence of positions 2 and 6, 2' and 6', or 3 and 5 and 3' and 5' is to be noted. R_1 and R_2 may be hydrogen, methyl, ethyl, propyl, isopropyl, butyl, isobutyl, pentyl, isopentyl, hexyl, isohexyl, a cyclohexyl including the methyl, ethyl, propyl butyl, pentyl and hexyl substituted cyclohexyl or a phenyl including the methyl, ethyl, propyl, butyl, pentyl and hexyl substituted phenyls. Diphenols of the types discussed when reacted with epichlorhydrin will produce diglycidyl ethers of the general formula:

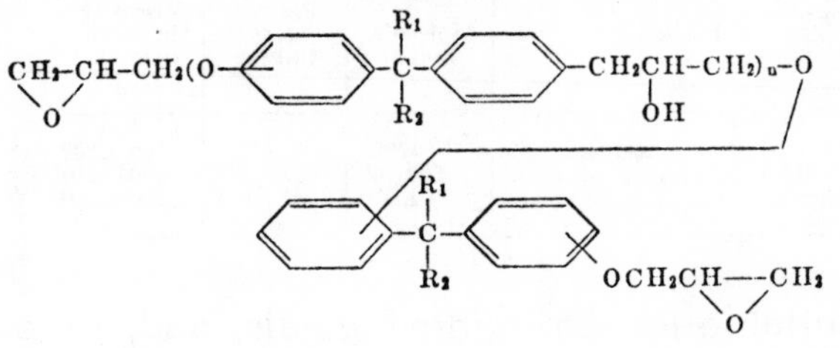

where R_1 and R_2 have the same designation as above, and epoxy propoxy groups are positioned as are the phenolic hydroxy groups in the parent diphenol. The value of n in such compositions is such as to give a fluid material and is preferably below 5. Preparation of such diglycidyl ethers from epichlorhydrin and the diphenols proceeds in the presence of a basic or alkali oxide, such as sodium hydroxide, described in U.S. Patent 2,506,486 noted above.

Mixtures of the polyepoxide and polysulfide compounds will react slowly in the absence of a catalyst, giving a soft gel within a period of several weeks. Inclusion of a catalyst, such as an alkaline substance will accelerate the reaction so that gelation time is considerably reduced. Suitable catalysts for such purposes are the polyamines or polyamides.

To effectively mask the offensive odor of the polysulfides contained in the polyepoxide compositions the proportions of styrene oxide as given above are employed. However, variations beyond the indicated limits may be tolerated without any deleterious effects.

Example 1: This example illustrates the masking of the odor of a liquid organic polysulfide polymer prepared as follows and available commercially from the Thiokol Corporation as "Thiokol LP-3."

2.5 mols of sodium hydrosulfide, as a 2 molar aqueous sodium hydrosulfide solution, and 2.5 mols of 2 molar sodium disulfide, as a 2 molar aqueous solution are thoroughly mixed. 25 g. of crystallized magnesium chloride dissolved in 50 cc of water are added to the mixture. The mixture is heated with agitation to a temperature of 160°F. and 4 mols of dichlorodiethyl formal containing 2% by weight of 1,2,3-trichloropropane are added dropwise over a period of 1 hour. The temperature of the mixture is not allowed to exceed 180°F.

After all the organic material has been added, the mixture is kept for an additional hour at 180°F. The resulting dispersion is allowed to settle, the supernatant liquid is decanted, and the remaining dispersion washed repeatedly with H_2O. The residuum is then acidified to a pH of about 6 to coagulate the material, and then is washed repeatedly with H_2O. 100 parts by weight of the resulting polymer and 5 parts by weight of styrene oxide were thoroughly stirred to blend the ingredients. The product was examined and the offensive polysulfide odor was not noticeable.

Example 2:   To illustrate the masking of the offensive odor of the above polysulfide when used in preparing an epoxy resin composition where the resin is a polyglycidyl ether of a phenol-formaldehyde condensation having a viscosity within the range of 6,000 to 16,000 centistokes and containing approximately 1 g. mol of epoxy group per 178 g.:

(a)   A mixture of 100 parts of the polyglycidyl ether, 25 parts of polysulfide, 10 parts of triethylene tetramine (TETA) curing agent and 10 parts of styrene oxide was prepared. The mixture was thoroughly stirred to blend the ingredients and maintained at room temperature. The product was examined and the offensive polysulfide odor was not noticeable.

(b)   The above procedure was repeated in the absence of styrene oxide. Examination of the composition clearly indicated the offensive odor of the polysulfide.

POLYSULFIDE POLYMERS MADE FROM MERCAPTO THIOPHOSPHITES AND -PHOSPHATES

Organic compounds containing phosphorus and sulfur are of considerable industrial importance and processes for preparing them are of considerable interest. Using a process developed by J.W. Stanley, Jr. (U.S. Patent 3,296,221; January 3, 1967; assigned to Phillips Petroleum Company), such phosphites are prepared by reacting a dimercaptan with phosphorus trichloride and recovering the phosphite from the resulting reaction mixture.

Such phosphites can be oxidized to obtain the corresponding phosphates. In addition, high molecular weight polysulfide polymers can be obtained by reacting the mercapto-substituted thiophosphites and -phosphates with dimercaptans in the presence of sulfur.

The dimercaptan which can be used to prepare the mercapto-substituted thio-phosphites and -phosphates and the corresponding polysulfide polymers can be represented by the general formula HS—R—SH, where R is an organo radical having 1 to 20 carbon atoms per molecule. R is preferably a radical selected from the group consisting of aliphatic, cycloaliphatic, and aromatic radicals, and combinations thereof. R can also be a heterocyclic radical such as:

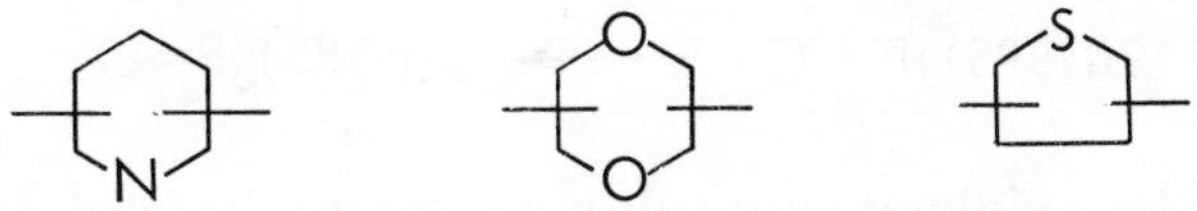

as well as other organic radicals.

The reaction of phosphorus trichloride with the dimercaptan to prepare the mercapto-substituted thiophosphites can be illustrated by the following equation:

$$PCl_3 + 3HS-R-SH \longrightarrow (HSRS)_3P + 3HCl$$

As shown in the equation, 3 mols of the dimercaptan are required to react stoichiometrically with the phosphorus trichloride and produce a mercapto-substituted thiophosphite having three free mercapto groups. The preparation of the mercapto-substituted thiophosphites are carried out in glass reaction vessels and equipment, though carbon steel equipment can be employed if a dry system is maintained, since the dry hydrogen chloride which is evolved is not corrosive. The reaction proceeds very rapidly above 60°F. and it seems best to carry out the initial reaction at a temperature of 80° to 150°F. at atmospheric pressure.

Temperature used should be low enough to prevent the loss of low boiling phosphorus tri-chloride (BP 169°F. at 760 mm.) with the evolved hydrogen chloride. Pressures above atmospheric pressure are not particularly suitable, since the hydrogen chloride formed should be removed to foster the reaction.

After an initial reaction period of 1 to 2 hours, using a nitrogen atmosphere or reflux conditions, reduced pressure and/or nitrogen stripping can be used to remove the evolved hydrogen chloride and complete the reaction. The use of a solvent or diluent other than the dimercaptan will not be required. The mercapto-substituted thiophosphites can be recovered from the reaction mixtures by any suitable method. The particular recovery method used will depend upon the particular dimercaptan reacted with the phosphorus trichloride, and the desired use of the phosphite product. For some purposes, e.g., where polysulfide polymers are to be prepared from the phosphite, the excess dimercaptan (unreacted) need not be removed; but where it is desired to remove the dimercaptan, it may be stripped off at reduced pressure to obtain a highly pure mercapto-substituted thiophosphite product. Also, where the phosphite is subsequently oxidized, for example with air, it will not always be necessary to remove the unreacted dimercaptan.

The mercapto-substituted thiophosphite product of this process can be used as an oxygen scavenger. For example, the phosphite can be incorporated into polysulfide polymers where it will take up the atmospheric oxygen entering the polymer and prevent the premature aging of the polymer.

Where a mercapto-substituted thiophosphate product is desired, it can be obtained by the simple oxidation of the phosphite precursor. For example, the mercapto-substituted thiophosphite can be blown with air at temperatures of 60° to 400°F., preferably 80° to 150°F., using at least 10 mols of oxygen per mol of phosphite to get at least 90% oxidation of the latter, as illustrated by the following equation:

$$2(HSRS)_3P + O_2 \longrightarrow 2(HSRS)_3P{=}O$$

The phosphites can also be oxidized with sulfur at temperatures of 280° to 310°F. and in the absence of catalytic systems such as those used in polysulfide formation, to prepare the corresponding mercapto-substituted tetrathiophosphate, as illustrated by the equation:

$$(HSRS)_3P + S \longrightarrow (HSRS)_3P{=}S$$

The mercapto-substituted thiophosphites and -phosphates can also be used as cross-linking agents in the preparation of branched chain polysulfide polymers, where the degree of branching within the polysulfide polymer is limited by the molar quantity of mercapto-substituted thiophosphite or -phosphate incorporated within the polymer.

These phosphites and phosphates are employed in the reaction of dimercaptans with sulfur, as illustrated by the following simplified equations:

$$4xHS-R-SH + 2x(HSRS)_3P + 5xS \longrightarrow \left[\begin{array}{c} H \left[\begin{array}{c} -SRS-SRSPSRS-SRS- \\ | \\ S \\ | \\ R \\ | \\ S \\ | \\ S \\ | \\ R \\ | \\ S \\ | \\ -SRS-SRSPSRS-SRS- \end{array} \right] H \end{array}\right]_x + 5xH_2S$$

$$4xHS-R-SH + 2x(HSRS)_3P{=}O + 5xS \longrightarrow \left[\begin{array}{c} H \left[\begin{array}{c} \overset{\overset{O}{\|}}{-SRS-SRSPSRS-SRS-} \\ | \\ S \\ | \\ R \\ | \\ S \\ | \\ S \\ | \\ R \\ | \\ S \\ | \\ \underset{\underset{O}{\|}}{-SRS-SRSPSRS-SRS-} \end{array} \right] H \end{array}\right]_x + 5xH_2S$$

Instead of elemental sulfur, sulfur-donor or sulfur-yielding compounds can be used in preparing the polysulfide polymers, and the term "sulfur" as used herein means such compounds unless qualified. The amount of sulfur which can be used in preparing the polysulfides can vary appreciably but will be usually about 2 to 5 mols of sulfur per mol of dimercaptan, excess sulfur being preferred.

The reaction between the dimercaptan, tri(mercaptoalkylthio)phosphite or phosphate, and sulfur will usually be carried out at temperatures in the range of 80° to 150°F. The particular temperatures used will depend upon the particular dimercaptan employed and the product desired, but generally temperatures in the range of 80° to 150°F. will be satisfactory. If the temperature is above 250°F., the phosphite product cannot be obtained as the sulfur will oxidize the phosphite to the phosphate.

The pressure at which the polysulfide polymer formation is carried out will usually depend upon the dimercaptan used, but generally will be sufficient to maintain the reactants in a liquid phase. Pressures of 0 to 15 psig will be sufficient. When superatmospheric pressures are required, an agent such as ZnO may be used to react with and remove the hydrogen sulfide gas generated by the aforementioned reaction. The reaction will vary appreciably but ordinarily will be from 0.25 to 10 hours, preferably 1 to 2 hours.

The formation of the polymers using sulfur is carried out in the presence of a catalyst, particularly those compounds having an alkaline reaction, with amines being preferred because of their solubility in the reaction mixtures.

The amount of catalyst used can vary appreciably, but preferably from 1 to 2 weight percent of the reaction mixture. In place of using sulfur in the formation of the polysulfide polymers, any catalyst or system capable of joining two mercapto groups can be such as di- or poly-valent metallic salt, oxide, or hydroxide, e.g., zinc hydroxide; or a dicarboxylic acid HOOC—R—COOH or a diacetyl halide XC(O)—R—C(O)X where R is an organo radical as defined above and X is a halo radical such as chlorine, bromine, and iodine; or air in the presence of cupric chloride and hydrochloric acid.

The process for the polysulfide polymers is carried out in the absence of a diluent. However, diluents suitable for use in the process are hydrocarbons which are not detrimental to the oxidation reaction. Also, straight and branched chain paraffins which contain up to and including 10 carbon atoms per molecule can be used.

The amount of diluent employed will vary appreciably and can range as high as 90 to 95% of the reaction mixture. These polysulfide polymers are generally rubbery in nature and can be used as caulking compounds, fuel sealing putty, oil resistant rubber, binders for solid rocket propellants and the like.

POLYSULFIDES WITH SILICON

Organosilicon Polysulfide Rubbers

A process developed by E.P. Plueddemann (U.S. Patent 3,317,461; May 2, 1967; assigned to Dow Corning Corporation) gives a substance composed essentially of a polymer of the formula:

$$\left[X_{(3-n)} \overset{\overset{\textstyle R''_n}{|}}{Si} \right]_m R'S[-CH_2(R)_b CH_2 S_a-]_c CH_2(R)_b CH_2 SR' \left[\overset{\overset{\textstyle R''_n}{|}}{Si} X_{(3-n)} \right]_m$$

b is an integer with a value of 0 to 1. R is free of aliphatic unsaturation, and is made up of divalent radicals made up of C, H, and optionally O and S atoms, the latter being in the form of $\equiv C-O-C\equiv$, $-OH$, $-SH$ and $\equiv C-S-C\equiv$ groups. Thus, R can be hydrocarbon, alcohol, mercaptan, ether or thioether divalent radicals.

a is an integer with a value of 2 through 4. The exact structure of the polysulfide group represented by the symbol S_a- can vary. Typical structures are $-S-S-$, $-S-S-S-S-$,

$$\overset{\overset{\textstyle S}{\|}}{-S-S-}, \quad \overset{\overset{\textstyle S \quad S}{\| \; \|}}{-S-S-}$$

and $-S-S-S-$. R' is a multivalent radical that can be composed entirely of C and H or it can contain oxygen in the form shown on the following page.

56

$$-\overset{\overset{\displaystyle O}{\|}}{C}-O-,\ \equiv C-O-C\equiv$$

or hydroxyl groups. Thus, R' can be a hydrocarbon ester, ether, or alcohol radical, but there can be no acetylenic unsaturation therein. The number of silyl groups attached to R' varies with the valence of R'. R" is a monovalent hydrocarbon or halohydrocarbon radical. X is a hydrolyzable radical, the favored ones being acyloxy, alkoxy and ketoxime

$$\left(\begin{array}{c} R \\ \diagdown \\ \diagup C=N-O- \quad \text{or} \quad \boxed{R\ \ C=N-O-} \\ R \end{array}\right).$$

n is an integer of 0 through 2 and m is an integer equal to one less than the valence of R'. c is an integer of value of at least one. There is no critical value for c but it is preferred that c have a value of at least 5 and a particularly desirable range is when c is from 5 through 100.

The term "consisting essentially of" as employed in the claims means that the polymers are substantially as shown but they can contain other groups in minor amount which do not change the basic character of the polymers. One such group is a branching group derived from a trifunctional halide as described infra. These can be incorporated into the polymers to give a branched structure which enhances the cure of the polymers. The compositions can be made by reacting polysulfide polymers of the formula:

$$HS[CH_2(R)_bCH_2S_a]_cCH_2(R)_bCH_2SH$$

with a silane containing a substitutent having the C=C group in the presence of peroxides or an alkaline catalyst such as alkali metal alkoxides. Peroxide catalysts are preferred for the addition of SH to nonconjugated olefins, while alkaline catalysts are most preferred for addition of SH to conjugated double bonds. The —SH group of the polysulfide adds across a double bond on a carbon substituent of the silane. The reaction is somewhat exothermic, and will go at room temperature or slightly above if a carbonyl group is conjugated with the double bond e.g.,

$$CH_2=CH\overset{\overset{\displaystyle O}{\|}}{C}-$$

Otherwise, heating of the reaction mixture is generally required. For instance, the following were mixed at room temperature:

50 g. of $\quad HS(CH_2CH_2OCH_2OCH_2CH_2S_2)_{\sim6}CH_2CH_2OCH_2OCH_2CH_2SH;$

45 g. of $\quad CH_2=\underset{\underset{\displaystyle CH_3}{|}}{C}-\overset{\overset{\displaystyle O}{\|}}{C}-OCH_2CH_2CH_2Si(OCH_3)_3$

and 1 g. of a 25% solution of sodium methoxide in methanol. Within 2 hours the typical polysulfide odor was replaced by a fruity ester odor. The product was an amber, viscous liquid that remained stable indefinitely if kept in an airtight container. The product was exposed to the air and the material turned very slowly into a solid. This reaction can be accelerated by adding specific catalysts for hydrolysis and condensation of methoxysilanes, such as stannous octoate.

Another way of making the materials is to react the sodium salt of the polysulfide polymer, (i.e., one having SNa groups in the place of the SH groups), with a silane that has a halogen atom on an organic substituent, for example:

$$\text{NaS(polysulfide polymer)SNa} + 2ClR'\text{(organosilane)} \longrightarrow$$
$$\text{(organosilane)R'S(polysulfide polymer)SR'(organosilane)} + 2NaCl$$

In general, heating is required for this reaction. A third method involves reacting the polysulfide polymer having terminal SH groups with an epoxy silane. The reaction can be represented by:

$$\text{HS(polysulfide)SH} + \overset{O}{\overset{/\backslash}{CH_2CH}}\text{--}\overset{R''_n}{\underset{|}{Si}}X_{3-n} \longrightarrow X_{3-n}\overset{R''_n}{\underset{|}{Si}}\text{----}\overset{\overset{H}{|}\,\overset{O}{|}}{CH}CH_2S\text{(polysulfide)}SCH_2\overset{\overset{H}{|}\,\overset{O}{|}}{CH}\text{--}\overset{R''_n}{\underset{|}{Si}}X_{3-n}$$

Examples of the process are given.

Example 1: The following were mixed and warmed to 120°C. 75 g. diallylitaconate, a trace of chloroplatinic acid, 0.1 g. phenyl-beta-naphthylamine, and 0.1 g. hydroquinone. To this was added 120 g. $HSi(OCH_3)_3$. The mixture was refluxed for 3 hours at 110° to 120°C. and finally stripped to 130°C. at 3 mm. Hg pressure. 100 g. of an oily residue was recovered, known as bis-trimethoxysilylpropylitaconate:

$$CH_2=\overset{\overset{O}{\|}}{\underset{\underset{H_2C-C-O(CH_2)_3Si(OCH_3)_3}{\underset{\|}{O}}}{C-C}}\text{-O(CH}_2)_3Si(OCH_3)_3$$

To 50 g. of this material was added 50 g. of

$$HS(CH_2CH_2OCH_2OCH_2CH_2S_2)_{23}CH_2CH_2OCH_2OCH_2CH_2SH$$

and 50 g. of toluene as a solvent. This was made alkaline with sodium methoxide in methanol and refluxed for 1 hour. The solution was then neutralized with $(CH_3)_3SiCl$, then filtered to give a clear amber liquid.

Thin films of this material were exposed to air for 24 hours. A sticky fluid resulted, indicating partial curing only. However, when a trace of $SnCl_4$ was added to the material before spreading it into a film, 24 hours of exposure to air caused it to cure to a flexible film.

Example 2: The following was warmed at 100°C. for 4 hours:

40 g. of $HS(CH_2CH_2OCH_2OCH_2CH_2SS)_{23}CH_2CH_2OCH_2OCH_2CH_2SH$

5 g. of

$$\overset{O}{\overset{\displaystyle \triangle}{CH_2CHCH_2OCH_2CH_2CH_2Si(OCH_3)_3}}$$

and 0.2 g. of

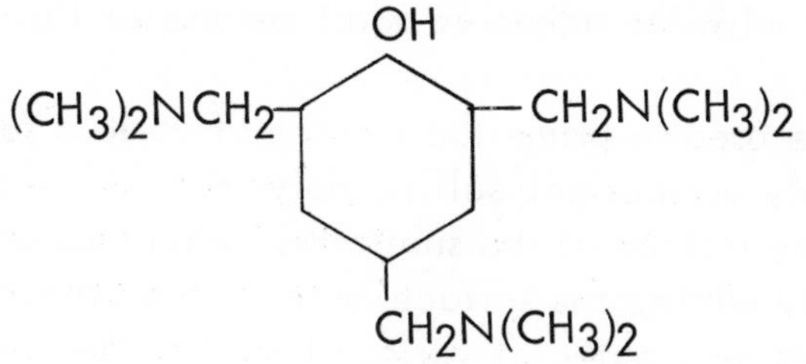

as a catalyst.

Reaction took place between the SH groups (and any incidental OH groups) on the polysulfide and the oxirane group on the silane to produce a compound of the average formula:

$$(CH_3O)_3Si(CH_2)_3OCH_2\overset{OH}{\underset{|}{C}}HCH_2S(polysulfide)SCH_2\overset{OH}{\underset{|}{C}}HCH_2O(CH_2)_3Si(OCH_3)$$

In these cases where the silane reacts with an OH on the polysulfide an O replaces the S in the above formula. The product was a viscous, fluid polymer with very little thiol odor. A portion of the condensed product was mixed with 0.1% by weight of isopropyl titanate and exposed as a film to the atmosphere. Within 16 hours the film had set to a clear, flexible, nontacky elastomer with excellent resilience.

Silane-Containing Sealant

A process developed by J.J. Giordano (U.S. Patent 3,312,669; April 4, 1967; assigned to Thiokol Chemical Corporation) gives a strong and lasting bonding of cured, liquid polysulfide polymer based sealant and caulking compositions to the substrate(s) being treated therewith, particularly in the presence of polar solvents, if the sealant composition is used in conjunction with one or more adhesive additive compounds having the structure $HR-R-Si-(OR')_3$ in which R and R' may be the same or different lower alkyl radicals such as methyl, ethyl, propyl, isopropyl, butyl, etc. In a given case, the three R' groups present in the compound(s) being used may be the same or different lower alkyl groups(s).

The adhesive additive compound may be used in the form of a prime coat wherewith the substrate to be treated is first primed and the liquid polysulfide polymer-based sealant compositions is then added to, and cured on, the thus-treated substrate.

The silicon-containing adhesive additive may also be used, according to this process, by being admixed into the curable, liquid polysulfide polymer-based sealant or caulking composition and then applying, and curing thereon, the curable composition to the substrate(s) being treated. Whether used as a priming agent and/or admixed in the sealant composition, the silane-containing adhesive additive should be used in such a quantity as to provide about

1 to 3.5 parts by weight of the additive per 100 parts by weight of the curable, liquid polymer being used. The adhesive additive may be termed a mercapto alkyl polyalkoxy silane. Examples of such compounds are mercapto propyl trimethoxysilane and mercapto propyl triethoxysilane.

The curable compositions have a pot life (working life) of 0.5 to 24 hours and the pot life of these compositions must be taken into consideration when admixing and/or bringing the curing agent in contact with the adhesive additive and/or liquid polysulfide polymer.

Adhesive additives were used to prime the surface of various substrates before an attempt was made to adhesively apply various polysulfide polymer-based sealant compositions. The additive was applied to the surface of the substrates being treated in the form of an ethyl alcohol solution by lightly wiping the surface once with a cotton swab saturated with the solution. The additive was evaluated at various levels of concentration in these solutions. The sealant formulations used were either a black or a white filled formulation having the following composition, expressed in parts by weight:

	White	Black
LP-32 polysulfide polymer	100	-
LP-2 polysulfide polymer	-	100
Witcarb RC (calcium carbonate)	30	-
Titanox RA 50 (titanium dioxide)	10	-
SRF #3 (carbon black)	-	30
Calcium stearate (thixotropic agent)	1	-
Stearic acid (retarder)	1	0.25
Aroclor 1254 (chlorinated hydrocarbon plasticizer containing 54% Cl)	5	5
HiSil 233 (hydrated silica)	3	2
Sulfur	0.1	-

Prior to their application to the primed substrate the sealant formulation was mixed with a lead peroxide and Aroclor 1254 curing paste admixture containing about 50% PbO_2. The priming solutions had the following composition, expressed in parts by weight:

Primary Solution

	1	2	3	4
Mercapto propyl trimethoxy silane	5	10	15	-
Mercapto propyl triethoxy silane	-	-	-	10
Ethyl alcohol	95	90	85	90

After the priming solution was applied to the substrate, it was allowed to stand overnight at room temperature during which time the solvent evaporated. After being applied to the primed surfaces, the sealant compositions were then allowed to cure for about a week at room temperature before the adhesion of the cured sealant to the substrate was tested. The tests were planned in such a way that two sets of identically coated substrates were provided so that the adhesion properties of the sealant to these substrates could be tested under two sets of conditions where the sealant was placed (1) at room temperature under tap water and (2) at room temperature while exposed to the atmosphere. The adhesion test consisted in attempting to peel a cured bead of the sealant composition from the substrate. If the bead peeled off the substrate easily without causing a rupture in the structure of the bead, this fact was noted as an adhesive failure. If the bead could not be readily removed from the substrate without tearing the bead apart, this fact was noted as cohesive failure. In the borderline instances, this fact was noted as a slight cohesive failure.

ISOBUTYLENE SULFIDE POLYMERS

A process developed by R.H. Gobran and S.W. Osborn (U.S. Patent 3,329,659; July 4, 1967; assigned to Thiokol Chemical Corporation) polymerizes isobutylene sulfide in the presence of a catalyst which is the reaction product of two components. One of these components is an organometallic compound of the formula R_2M where R is alkyl or aryl and M is a metal of Group IIb of the periodic system, i.e., zinc, cadmium or mercury.

The second component, which is reacted with the organometallic component to form the catalyst can be generally characterized as a substance having at least one pair of unshared electrons. However, all substances falling within this broad genus do not appear to be operative. One relatively large sub-genus that has been found to be operative comprises compounds having an active hydrogen atom.

In addition to this sub-genus, i.e., the compounds having an active hydrogen, it has been found that useful catalysts can be prepared employing as the second component or cocatalyst elemental oxygen or sulfur, carbonyl sulfide and carbon disulfide. The preferred catalyst is the reaction product of diethyl zinc and water.

The resulting polymers are solid, highly crystalline materials of high molecular weight which have melting points in the range of 160° to 195°C. or more. The polymers can be molded into a variety of items using any of the molding processes conventionally used for thermo-plastic resins. Polymers having the lower melting points, i.e., below about 175°C. have lower molecular weights and are more soluble in organic solvents generally than the poly-mers having higher melting points, i.e., above about 175°C. The preparation of isobutylene sulfide is disclosed at pages 1050-1052 of the Journal of the Chemical Society, 1946.

In general, the polymerization process is executed by bringing the monomeric isobutylene sul-fide into contact with the catalyst prepared in the general manner described above. The cat-alyst is used to the extent of about 0.1 to 1%, based on the weight of the monomeric mate-rial being polymerized. The polymerization can be carried out satisfactorily at temperatures of 0° to 80°C. and is preferably conducted at 10° to 35°C. It has been found that polymers having the higher molecular weights and melting points are more readily prepared at reaction temperatures of 0° to 10°C. The reaction may take several hours to several days.

The polymerization reaction may be conducted without using a solvent, or it may be conducted in inert organic solvents such as aromatic hydrocarbons. The reaction mixture may be agitated to facilitate the reaction. The pressure at which the reaction is carried out does not appear to be particularly critical. Thus the reaction may be conducted in an open vessel at atmospheric pressure or in a closed vessel under autogenous pressure. In a modification of the process where a solvent is used, the reaction vessel is desirably charged with the solvent, catalyst and monomer in that order. Where no solvent is used, however, it is preferable to preform the catalyst and then react the preformed catalyst with the monomer. The reaction system should also be anhydrous except for the amount of water required to form the diethyl zinc-water catalyst as described previously.

One and a half milliliters of a solution of diethyl zinc in benzene containing 4.75 mmols of diethyl zinc per milliliter of solution was dissolved in 50 ml. of dry tetrahydrofuran. To 41.5 ml. of this solution, there was then added 0.082 ml. of distilled water. The resulting solution thus contained a molar ratio of 0.75 to 1 of water to diethyl zinc.

A clean, dry 12 ounce Coca-Cola bottle was sequentially charged with 11 g. of isobutylene sulfide, 11 ml. of sodium dry benzene and 4.7 ml. of the catalyst solution prepared above. This catalyst solution contained 0.282 mmol of diethyl zinc and 0.211 mmol of water. As soon as the catalyst had been added, the mixture in the Coca-Cola bottle exothermed to 55° to 60°C. After about one-half hour, the temperature of the system had subsided to about room temperature.

The bottle was then sparged with nitrogen and sealed. It was allowed to remain at room temperature for about 5 1/2 days. The bottle was then opened and the contents were poured into 250 ml. of petroleum ether. The polyisobutylene sulfide which had been formed readily precipitated and was washed 3 times with 250 ml. portions of petroleum ether. The polymer product was then filtered and dried to constant weight. A yield of 4.2 g. or 38% of theoretical was obtained. The polymer had a MP of 185°C. It was a crystalline white powder which is moldable into articles having various shapes and sizes using conventional molding procedures.

ISOCYANATE-TERMINATED POLYMERS

A process developed by A.F. Santaniello (U.S. Patent 3,386,963; June 4, 1968; assigned to Thiokol Chemical Corporation) gives storage-stable isocyanate-terminated polysulfide polymers that are curable at room temperature by atmospheric moisture, with or without additional curing agents or catalysts, to elastomers having good properties for a variety of applications. The polymers are believed to have the structure:

$$OCN-[R-NH-\underset{\underset{O}{\|}}{C}-O-R'SS(R''SS)_nR'-O-\underset{\underset{O}{\|}}{C}-NH]_m-R-NCO$$

in which R is an alkylene or arylene radical, R' and R" are the same bivalent aliphatic radical where the carbon chain may contain oxygen atoms, n is 1 to 100 and m is 1 or more, preferably up to 15. The polymers can be made by reacting a polyisocyanate of the formula, $R(NCO)_x$ in which x is at least 2 with a polymer of the formula $HO-R'SS(R''SS)_nR'-OH$ at ratios of NCO:OH greater than 1.0, preferably up to 4.

Example 1: In the process, a reaction vessel was charged with 161.2 g. (0.925 mol) of an 80/20 mixture, respectively, of the 2,4- and 2,6-isomers of toluene diisocyanate. To the vessel were added 430 g. (0.421 mol) of a hydroxyl-containing polysulfide polymer having essentially the structure:

$$HO-C_2H_4OCH_2OC_2H_4(SS-C_2H_4OCH_2OC_2H_4)_5SS-C_2H_4OCH_2OC_2H_4-OH$$

over a period of about 10 minutes. The reaction mixture was heated to about 100°C. in 15 minutes, maintained at 100° to 120°C. for about 50 minutes, and then cooled to 60°C. and stored in a glass jar. The isocyanate-terminated polymer thus prepared had an —NCO content of about 6.9% as shown by infrared analysis, and was formulated as shown in the following table into compositions from which castings were made.

Formulation	A	B	C	D	E
Polymer (g.)	64.10	63.18	61.95	64.10	64.10
Triisopropanolamine (g.)	6.375	6.738	4.403	3.188	0.3188
Polypropylene glycol 425 (g.)	-	-	-	2.1256	4.038
Tack-Free Time at 120° ±2°C. (min.)	10	10	8	50	90
Total Cure Time at 120°C. (min.)	130	130	128	170	210

Physical Properties After an 11 Day Aging Period

Formulation	A	B	C	D	E
Tensile strength (psi)	762	854	753	-	648
Elongation (percent)	92.5	80	80	-	77.5
Shore "A" hardness	83	82	84	-	62

Example 2: A reaction vessel containing 60.95 g. (0.35 mol) of the toluene diisocyanate mixture described in Example 1 was charged with 261.6 g. (0.1 mol) of a hydroxyl-containing polysulfide polymer having essentially the structure:

$$HO-C_2H_4OC_2H_4-(SS-C_2H_4CC_2H_4)_{17}SS-C_2H_4OC_2H_4-OH$$

over a 17 minute period. The reaction mixture was heated to 80°C. over a 20 minute period, maintained at 80° ± 3°C. for 2 hours and then allowed to cool to room temperature. The isocyanate-terminated polymer thus prepared contained about 8.3% —NCO as shown by hydrochloric acid titration. Castings were made from the formulations of the polymer shown in the table below.

Formulation	A	B	C	D	E
Polymer (g.)	34.8	34.8	34.8	34.8	34.8
Triisopropanolamine (g.)	3.48	-	-	-	-
Hexanetriol (g.)	-	2.68	-	-	-
Triethanolamine (g.)	-	-	2.98	-	-

(continued)

Formulation	A	B	C	D	E
Phenyldiethanolamine (g.)	-	-	-	5.34	-
Polysulfide polymer of example 2 (g.)	-	-	-	-	39.24
Tack-free time at about 120°C. (min.)	115	*	103	100	90
Total cure time at about 120°C. (min.)	235	232	223	220	210

*Surface cure.

Example 3: The process can also utilize a mercaptan-terminated polysulfide polymer. For example, to a five liter one-neck round bottom flask are added 3,000 g. of a mercaptan-terminated polysulfide polymer of the formula:

$$HS-C_2H_4OCH_2OC_2H_4-(SS-C_2H_4OCH_2OC_2H_4)_6-SH$$

1,500 ml. of m-xylene and 4.5 g. of p-toluene sulfonic acid dissolved in 10 ml. of water. A condenser is inserted into the neck and the mixture is heated at reflux for 24 hours. Silver nitrate titration is used to determine the absence of thiol groups, and thereafter the catalyst is neutralized with CaO. The xylene is removed under vacuum in a Roto-drier in which the water is also removed as part of a azeotrope. An essentially quantitative yield of the corresponding hydroxyl-terminated polysulfide polymer is obtained.

100 parts (0.086 mol) of the polymer thus prepared and 10 parts of benzene are added to a flask where the mixture is slowly agitated. The flask is swept with dry nitrogen and a dry nitrogen atmosphere is maintained throughout the remainder of the process. The mixture is heated to its boiling point at about 70°C., and benzene and water are removed by slow azeotropic distillation.

After removal of all the benzene and water, the mixture is cooled to about room temperature and about 22.5 parts (0.129 mol) of an 80/20 mixture of 2,4- and 2,6-toluene diisocyanate, respectively, are added to the flask. Under constant agitation, the mixture is heated to 80°C. and held at that temperature for 3 hours. The mixture is then cooled to about room temperature and stored in glass jars under a nitrogen atmosphere. An isocyanate-terminated polymer is thus prepared from reactants having an NCO/OH ratio of 1.50.

POLYSULFIDES PLUS SULFUR

A process developed by L.E.A. Godfrey and J.A. Garman (U.S. Patent 3,427,292; February 11, 1969; assigned to FMC Corporation) mixes elemental sulfur with an organic polysulfide selected from the group consisting of bis(dialkoxyphosphinothioyl)polysulfides and bis(alkoxythiocarbonyl)polysulfide of the following formulas:

$$ROC-(S)_n-COR' \quad and \quad (RO)_2P-(S)_n-P(OR')_2$$

where R and R' which may be alike or different represent alkyl radicals of 1 to 18 carbon atoms which in the case of the lower members, i.e., 1 to 4 carbon atoms, can be substituted by phenyl or naphthyl, and n is an integer of from 2 to 4.

A number of 5.0 g. quantities of insoluble sulfur were weighed into crystallizing dishes and the samples mixed with a solution of the polysulfide. The solutions contained 0.025 g. of the polysulfide in 15 to 25 ml. of carbon disulfide. A like number of samples of insoluble sulfur were weighed out but omitting the polysulfide. These untreated samples serve as controls or blanks. The solvent was allowed to evaporate at room temperature after which the insoluble sulfur and polysulfide mixtures were dried 30 minutes in a vacuum oven at 40°C. at 250 mm. After the insoluble sulfur had been thoroughly mixed with a spatula the material was again assayed. The samples were then divided into equal parts and heated in a constant temperature oven for varying lengths of time. At the end of the heating period samples were again assayed.

In a variation, ordinary sulfur was thoroughly mixed with 5.0% of bis(n-butoxythiocarbonyl) disulfide by heating the components to 150°C. Except for a relatively small residue, the additive appeared to be dissolved in the molten sulfur. The liquid was poured into a Pyrex glass mold and allowed to cool. Subsequently the cast rod was removed from the mold and its impact strength measured by the Izod test. For comparison purposes control castings were prepared not containing the additive.

The results show a significant increase in impact strength. Cast samples containing bis(n-butoxythiocarbonyl)disulfide were also found to be less friable than pure elemental sulfur. Such compositions are excellent road marking materials. The organic polysulfide is used in relatively minor proportions compared with the sulfur substrate. Generally speaking, the compositions are preferably formulated containing from 0.01 to 5.0% by weight of the organic polysulfide.

DICHLOROALICYCLIC SULFIDE POLYMERS

Synthetic polyolefins such as polystyrene, etc., form the basis for the predominance of the plastic articles now being manufactured. The polyolefins, particularly polystyrene, have achieved this wide spread utility primarily because of their comparatively low cost, adequate mechanical characteristics, and their ability to be molded, etc.

A process developed by R.A. Meyers and E.R. Wilson (U.S. Patent 3,429,859; February 25, 1969; assigned to TRW, Inc.) gives poly(dichloroalicyclic sulfide)polymers that have improved mechanical properties and may be used in preparing coatings, laminates, adhesives and various molded articles.

The dichloroalicyclic sulfide and disulfide polymers are prepared by reacting a diunsaturated alicyclic compound selected from the group consisting of cyclopentadiene, dicyclopentadiene, and a halogen or lower alkyl substituted cyclopentadiene or dicyclopentadiene with a sulfur chloride to obtain linear dichloroalicyclic sulfide and disulfide polymers with an average molecular weight of at least 10,000. The cyclopentadienes are reacted on a molar basis from about −20°C. up to about room temperature.

The amount of sulfur chloride, either sulfur monochloride or sulfur dichloride which may be chemically reacted with the cyclopentadienes to form the corresponding polymers is proportional to the unsaturation of the diene, i.e., the number of double bonds per unit weight which is approximately one mol of sulfur chloride for each mol of mono or dicyclopentadiene.

The reaction temperature most suitable for the preparation of the polymers is about $-20°C.$, but the temperature may range to about room temperature or above for periods of from 2 to 24 hours. In any event, the reaction temperature and time is maintained to insure a reaction between the cyclodiene and the sulfur chloride so as to obtain addition of the sulfur and chlorine atoms to the double bonds of the ring and to insure a high molecular weight.

In addition, it has been found to be advantageous to add to the reaction mixture a suitable diluent which is nonreactive with the sulfur chloride. When the reaction is carried out in a closed vessel, 5 to 100% by weight of a low boiling solvent including the paraffinic, naphthenic, or aromatic solvents may be used to control the reaction.

In carrying out the synthesis, it is desirable to mix the diene in one portion of the solvent, e.g., approximately 80% of the total amount and the sulfur chloride in the remaining portion. Subsequently, the diene solution, in a closed flask, is cooled to approximately $-20°C.$ and the sulfur chloride solution is added dropwise while the temperature is maintained. The final polymers depending upon the amount of sulfur chloride added to the unsaturation of the dienes may be characterized as hard thermoplastic resins with softening points of about $60°C.$, and in some cases may range as high as $200°C.$ as determined by the ASTM ring and ball method depending upon the particular reactants and conditions employed.

In detail, the process with cyclopentadiene follows this route. To a solution of 67.3 parts by weight (1.02 M) of cyclopentadiene in 300 parts by weight of methylene chloride in a one liter, three-necked, round bottom flask which was equipped with a thermometer, dropping funnel, and a magnetic stirrer, was added a solution of 107.8 parts by weight (1.02 M) of freshly distilled sulfur dichloride in 300 parts by weight of methylene dichloride over a period of one hour while the temperature was maintained at about $-20°C.$ by means of a Dry Ice-acetone bath.

When the addition was completed, the reaction vessel was transferred to a freezer and maintained at about $-20°C.$ for about 24 hours. At the end of this time, the reaction mixture was allowed to warm to room temperature and the solvent was removed by vacuum stripping. The product was characterized as a yellow to yellow-brown polymer which softens at about $35°C.$ The polymer was then recoagulated from methylene chloride-petroleum ether with a 66% yield of a yellow-white polymer having a softening point of $65°C.$

The total product yield was approximately 169.3 parts by weight (96.7%). The inherent viscosity was 0.30 (0.5% solution in dimethylacetamide). A thermagravimetric analysis gave a residue of 35% at $800°C.$ with the first break occurring at $200°C.$ The elemental analysis of the product was as follows:

Carbon	35.92%	Sulfur	18.38%
Hydrogen	3.93%	Chlorine	39.20%

The theoretical chemical analysis of the polymer $(C_5H_6SCl_2)_n$ was calculated as carbon 35.52%, hydrogen 3.58%, sulfur 18.96%, and chlorine 41.94%. The infrared spectrum showed the absence of double bonds, as expected, for the polymer.

CURING

CURING UNDER ALKALINE CONDITIONS

An early curing process, developed by W.S. Foulks (U.S. Patent 2,584,264; February 5, 1952; assigned to Thiokol Corporation) deals with preparing a polymer having a molecular weight of about 1,200 and curing it with an alkaline agent (ammonia, for example). In this particular example, the polymer will be prepared by the oxidation of dimercapto diethyl formal or: $HSCH_2CH_2OCH_2OCH_2CH_2SH$.

Seven and one half mols of dimercapto diethyl formal are dispersed in water to which is added 14 mols of sodium hydroxide and the dispersion so formed is carefully protected from atmospheric oxidation. To the dispersion, while rapidly being stirred, is added a solution of hydrogen peroxide containing 6.5 mols of H_2O_2 and therefore equivalent to 6.5 atoms of available oxygen. The reaction takes place at room temperature and, after the addition of the hydrogen peroxide solution, is continued for about 15 minutes using a water bath when necessary to prevent the temperature from rising above about 60°C. At the end of the reaction period, the reaction liquid is treated with a solution of acetic acid containing 15 mols of acetic acid or slightly more than the equivalent required to neutralize the reaction liquid which is then allowed to settle and the supernatant liquid poured off. The oily layer is then purified by successive washings with water and intermittent settling and decantation.

Since the product is a polymercaptan, it is responsive to oxidation under alkaline conditions. The polythiopolymercaptan may be applied to a surface as a film. The alkaline condition is supplied by exposing the film to an alkaline substance in the vapor phase, e.g., ammonia or volatile amines. The oxidizing condition may be supplied by air, oxygen, or other gaseous oxidizing agent or preferably by a liquid or solid oxidizing agent mixed with the polymer prior to filming it. It is the oxidizing function that is important rather than the specific nature of any particular oxidizing agent, in conjunction with alkaline conditions. Mixtures of the polymer and certain oxidizing agents are stable when anhydrous. They may be stored and transported and become activated upon exposure to moist alkaline conditions.

The polymer produced is filmed out on any surface which it is decided to coat and the film thus obtained is treated with a volatile alkaline material such as ammonia. The polymerization takes place as a result of the action of the alkaline material in the presence of the moisture and oxygen of the air and the complete polymerization is usually finished over a period of from five to seven hours at room temperature.

USE OF METAL SALT OF 8-HYDROXYQUINOLINE

Polysulfide polyers exhibit a number of decidedly objectionable properties, one of which is known as cold flow. They undergo deformation under pressure with but slight recovery. While the products undergo a transformation from a plastic to an elastic state so essential for practical use, nevertheless the transformation is not complete enough for many applications. The vulcanization is generally effected with metallic substances, ordinarily zinc oxide, in contrast to sulfur the curative normally used for rubber.

Another disadvantage of polymeric organic sulfide elastomers is that they support fungus growths and are, therefore, objectionable for making sanitary garments and other objects in which it is desirable to have an elastomer having marked antiseptic, fungistatic or therapeutic properties.

A process developed by P.G. Benignus (U.S. Patent 2,588,796; March 11, 1952; assigned to Monsanto Chemical Company) involves the addition to the polymeric organic sulfide of a metal salt of 8-hydroxyquinoline. Examples of typical metals salts which may be used include copper, zinc, iron, magnesium, manganese, cadmium and aluminum salts. While as little as 0.1% by weight on the elastomer exhibits an appreciable effect, it is preferred that 1.5 to 3.5% by weight on the elastomer of a metal salt of 8-hydroxyquinoline be utilized. Amounts up to 5% by weight may be used to advantage although amounts in excess of 5% generally show no increased benefit. The curing process may be applied to any of the polymeric materials obtained by the reaction of inorganic polysulfides with organic dihalides.

Example: A stock containing about 2% of the preferred adjuvants was compounded from an elastomer comprising the polymeric reaction product of sodium polysulfide and an organic dihalide and compared to a similar stock containing no hydroquinoline salt. The composition of the stocks was as follows:

	Stock—Parts by weight	
	A	B
Elastomer	100.	100.
Carbon black	40.	40.
Zinc oxide	6.	6.
Dithio bis benzothiazole	0.3	0.3
Diphenyl guanidine	0.1	0.1
Stearic acid	0.5	0.5
Copper salt of 8-hydroxyquinoline	--------	3.04

The stocks so compounded were cured by heating in the usual manner in a press for 10 minutes at 310°F. Samples of the cured products were artificially aged by heating in an oven for 7 days at 158° to 160°F. The physical properties of the cured products before and after aging are set forth in the table below:

	Stock	Tensile at Break in lbs./in.^{3}	Ultimate Elongation	Shore Hardness
Unaged	A	1,150	442	59
Do	B	1,387	512	62
Aged	A	1,242	395	67
Do	B	1,743	433	73

The data show that the B stock containing the copper salt of the oxyquinoline had superior physical properties. It will be noted that the tensile strength was high and although the hardness was increased, the ultimate elongation was also higher. The resistance to cold flow was markedly enhanced by the presence of the copper quinolinolate.

The accelerating and curing activity of copper quinolinolate was found to be considerable. For example, the zinc oxide level commonly used in polymeric organic sulfide compounds is 10 parts per 100 parts of elastomer but it is possible and, in fact, preferable to employ lesser amounts in conjunction with the new adjuvants. For example, the vulcanizing effect of copper quinolinolate is so marked that with the usual amounts of zinc oxide there is a tendency to vulcanize during processing. The fungistatic qualities of the stocks were examined by standard procedures. The results of these tests showed that while the A stock supported fungus growth in every case, the B stock exhibited marked retardation of growth.

CROSS-LINKING COMPOUNDS

In a process developed by T.A. Te Grotenhuis (U.S. Patent 2,631,994; March 17, 1953; assigned to The General Tire and Rubber Company), rubber-like polymers of improved resistance to cold flow after vulcanization are obtained by reacting an alkaline polysulfide, which has at least two and preferably about three to five sulfur atoms per molecule, with a mixture of one or more saturated base compounds and one or more "substitute" compounds or cross-linking compounds, i.e., organic compounds containing at least one hydroxyl or amino group in addition to the substituents which are affected by reaction with the inorganic polysulfide. The base compounds (this term is herein used to designate compounds having an absence of groups for cross-linking) are preferably entirely saturated and are present in major proportion and have two, and preferably only two substituents such as halogen which are capable of being split off during the polymer-forming reaction.

These substituents are connected to different connected carbons. The substituted or cross-linking compounds (having a hydroxyl and/or an amino group for cross-linking) are present in only minor proportions to provide for cross-linkage during vulcanization with agents effecting cross-linkage between the substituted (hydroxyl and/or amino) groups. Since these groups apparently may be in large measure reacted during vulcanization with dicarboxylic acid, anhydride or chloride, or di-isocyanate, there is no appreciable decrease in chemical resistance of the rubbery material. The substituted or cross-linking compounds like the base compound or compounds contain two groups, such as halogen, which are split off during the polymer-forming reaction.

Example 1: One version of the process consists of dissolving 500 grams of hydrated sodium sulfide in about 1 liter of water and the solution boiled with 200 grams sulfur. The polysulfide formed, which has the formula Na_2S_{4-5}, is diluted with water to have a specific gravity of about 1.25. About 250 grams of ethylene dichloride and about 7 grams of glycerol dichlorohydrin (the molar equivalent of about 6 or 7 grams of butadiene dichloride) are intimately mixed with about 100 grams of ethyl alcohol and the mixture·is gradually added to the polysulfide solution with vigorous stirring. This mixture is heated at about 70° or 80°C. in a vessel having a reflux condenser, the reaction being completed in about two hours.

The plastic product remaining after the withdrawal of the liquid is washed with water. After drying, it is mixed with 60 parts of carbon black (based on 100 parts of the rubber plastic), about 8 parts of zinc oxide, and about 5 grams of phthalic acid anhydride (about that required to completely react with the hydroxyl groups of the 7 grams of glycerol dichlorohydrin). The compounded material when cured has substantially no tendency for cold flow and has excellent resistance to chemicals.

Example 2:   In another method, 150 parts of ethylene dichloride are mixed with about 10 parts of propylene dichloride and 25 parts of glycerol dichlorohydrin. The mixture is added slowly, with stirring, to a dispersion of about 260 parts of sodium polysulfide in 1,100 parts of water, containing about 5 parts of a stabilizing agent such as sodium alkyl naphthalene sulfonate, and about 12 parts of freshly precipitated magnesium hydroxide. The temperature of the reaction mixture is maintained at around 70°C. with constant stirring during the first part of the reaction and gradually raised, until 90°C. is reached. The latex is coagulated in the usual manner with salt and sulfuric acid, and the coagulum dried to obtain a rubbery product. The rubber product is compounded in accordance with the following formula, in which the parts are by weight:

	Parts
Rubbery polymer	100
Zinc oxide	5
Carbon black	60
Diphenyl guanidine	0.1
Maleic anhydride	4

The above ingredients are mixed in usual order on a cold mill; the maleic anhydride, which serves as a cross-linking agent as above described, being added last. The product is cured at the usual rubber vulcanizing temperature of 300°F. for about fifteen minutes. The cured product exhibits very little cold flow compared to the usual cured olefin polysulfide rubber products.

While the base compound is preferably entirely saturated and inasmuch as unsaturation (double bonds) apparently does not enter into any cross-linking reaction with the dicarboxylic or cross-linking agents, etc., but merely decreases chemical resistance of the polymer, the base compound may contain unsaturated groups or may be a mixture of saturated and unsaturated dihalides, etc.

In the above examples, the cross-linking agent used may be substituted in whole or in part by a molar equivalent of any of the above isocyanates, or above-designated acid anhydrides, or acids themselves with substantially equivalent results.

USE OF IODINE IN CURE

A process developed by W.L.A. Barth (U.S. Patent 2,727,883; December 20, 1955; assigned to Aero Service Corporation) provides a method of inducing polymerization of polysulfides which enables the curing time to be properly regulated with relation to the working life and the total of working life and curing time confined within relatively definite predetermined and practical limits, thus facilitating production from such polysulfides of molded objects and the like more economically and satisfactorily than has been possible.

In the polymerization of LP-2, one of the Thiokol rubbers, it has been the practice to employ zinc sulfide as a pigment, dibutyl phthalate as a plasticizer, cumene hydroperoxide as a catalyst and diphenylguanidine as an activator rendering the mass alkaline which is essential to the reaction when cumene hydroperoxide acts as a catalyst as well as an oxidizing agent. It has been found, however, that when the working life is extended by appropriate proportioning of these ingredients the curing time is likewise extended sometimes to 20 to 40 times as long as the working life and efforts to obtain a working life of two hours or more sometimes result in mixtures which cannot be cured or cannot be cured within any reasonably practical period of time.

As considerations of efficiency dictate that the curing time be made as short as possible whether the working life be long or short, it is impractical to use the reagents just mentioned when a relatively long working life is required and the utility of polymerizable polysulfides of the class here under discussion has heretofore been limited by that fact.

When a suitable iodine donor is added to LP-2 in the presence of other reagents, the working life can be controlled and extended, even to more than 6 hours if required, without unduly extending the curing time which for practical purposes it is considered should not exceed four days since if the curing time be prolonged beyond this limit an unsatisfactory product may result.

More specifically, when there is added to 100 parts by weight of LP-2 about 26.5 parts of dibutyl phthalate, zinc sulfide 46.70 parts, elemental iodine in a 0.5% solution in dibutyl phthalate 0.03 part, cumene hydroperoxide 8.25 parts and 0.22 part of a suitable antifoaming compound such as that available in the market under the trade name "Dow-Corning antifoam" supplied in 50% solution in benzol, a working life of 3 hours and 50 minutes is attained and on standing overnight the mass is found to have jelled by morning while at the expiration of two days after pouring, the fully polymerized material may be removed from the mold and subjected to whatever finishing treatment may be desired, this treatment, in the manufacture of relief maps, usually including surface lettering, coloring and the like. As the iodine donor, compounds of iodine as distinguished from the element itself in solution can be used.

USE OF CHROMATES

Chromic Acid Salts

A process developed by G. Gregory and I.P. Seegman (U.S. Patent 2,787,608; April 2, 1957; assigned to Products Research Company) uses, in the cure of polyalkylene polysulfides, a group of soluble salts of chromic acid such as chromates and bichromates including the sodium, patassium and ammonium chromate and bichromate salts. Any soluble salt of chromic acid which, upon solution, liberates anions containing chromium may be used in carrying out the process provided it has a solubility in the solvent equal to or greater than that of the potassium bichromate. Since the solubility of potassium bichromate at 20°C. is indicated in standard reference texts to be about 12 grams per 100 ml. of water (Solubilities of Inorganic and Organic Compounds, Volume I, page 528, Seidell, 1919 ed.), any salt of chromic acid having an equivalent or greater solubility and the ability to liberate chromium anions may be successfully employed.

These curing agents may be incorporated in a polyalkylene polysulfide in the form of finely divided solids or in the form of a solution. In the event the base mixture is to contain fillers or additives, such fillers or additives may be blended into the liquid polymer on a suitable mill such as a roller mill or paint mill and the curing agent thereafter incorporated into the mixture in any suitable manner. The curing agents are added to the polyalkylene polysulfides in stoichiometric proportions although this is not critical. In general, the curing agents are added at the rate of between about 3 and 10 parts by weight per 100 parts of the polyalkylene polysulfide, less than 3 parts may undesirably retard the rate of cure or prevent complete vulcanization whereas an excess (for example, 20 or 25 parts per 100 parts of the polymer) may tend to impair desirable properties and increase the cost. A cured composition resulting from the use of polymers with a curing agent alone may contain as high as 97% by weight of such polymers.

These curing agents perform the cure at normal atmospheric temperatures without appreciable exothermic reaction and can cure in a void without need of outside air, at a rate sufficiently slow to permit a batch of composition to have a reasonably long application life (that is, period of time after mixing during which the batch can be extruded, cast, molded or otherwise applied in use) without causing appreciable curing shrinkage or undue expansion, and without impairing either physical properties. Moreover, the curing agents are compatible with the presence of fillers, retarders, adhesion additives and other components of a batch.

Chromates with Hydrous Salts

A process developed by D.J. Smith (U.S. Patent 2,940,958; June 14, 1960; assigned to Thiokol Chemical Corporation) uses water-activatable oxidizing agent and hydrous salt. The water-activable oxidizing agent may be a metal peroxide or a chromate. The hydrous salt component of the curing composition should be stable at ambient temperatures and be capable of releasing water of hydration at temperatures between about 100° and 220°F. Such salts as sodium tetraborate $Na_2B_4O_7 \cdot 10H_2O$, and calcium ferrocyanide $Ca_2Fe(CN)_6 \cdot 12H_2O$, are outstanding.

Example: Samples of two representative liquid polysulfides were each admixed with preselected proportions by weight of ammonium dichromate and sodium tetraborate. The admixed samples were tested for storability at room temperature and for effectiveness of cure at temperatures of 140° and 220°F. One of the polysulfide samples (sample A) had an average structure represented by the formula:

$$HS(C_2H_4OCH_2OC_2H_4SS)_{23} - C_2H_4OCH_2OC_2H_4SH$$

occasionally having a side mercaptan group in the chain or recurring units. The other (sample B) had the average structure represented by the formula:

$$HS(C_2H_4OCH_2OC_2H_4SS)_6 - C_2H_4OCH_2OC_2H_4SH$$

These polysulfide samples are thus of a class which are essentially polythiopolymercaptan polymers, that is, polymers having as recurrent units saturated aliphatic oxahydrocarbon radicals interconnected by polysulfide groups. The conditions and results of these tests are summarized in the table below.

Sample A, parts by weight	100	100	100	--
Sample B, parts by weight	--	--	--	100
$(NH_4)_2Cr_2O_7$	10	--	10	20
$Na_2CrO_4 \cdot 4H_2O$	--	15	--	--
$Na_2B_4O_7 \cdot 10H_2O$	--	--	3	5
Remarks:				
48 hrs. at room temp.	No cure	No cure	No cure	No cure
24 hrs. at 140°F.	No cure	Cured	Cured	Cured
24 hrs. at 220°F.	No cure	Cured	Cured	Cured

These data indicate that both liquid polysulfides remain uncured at room temperature with and without the presence of hydrous salt and that the polysulfides containing ammonium dichromate and sodium tetraborate are cured by being subjected to temperature of 140°F. or more for twenty-four hours. These data also indicate the remarkable ability of hydrous sodium chromate to perform the functions of both the water-activatable curing agent and of the water-releasing hydrous salt. Many modifications involving time, temperature, nature of the chromate, and the type of hydrous salt are possible.

Faster Chromate Cure

When cured in the manner described previously in U.S. Patent 2,787,608 (employing as a curing component a soluble salt of chromic acid), cured elastomers may be produced having enhanced physical and chemical properties, e.g., excellent resistance to a wide range of solvents, oils and fuels; a desirable resiliency; a service temperature range of from about -70° to 350°F.; excellent ozone and oxidation resistance; and good electrical properties and adherence to metal at ordinary temperatures. Notwithstanding these desired characteristics and the demonstrated usefulness of the liquid polyalkylene polysulfides cured in the above manner, the cured elastomers are nevertheless subject to certain short comings:

(1) Their resistance to swelling and loss of physical properties after emersion in water at elevated temperatures is limited and exists only for relatively short periods of time, yet such conditions of moisture and temperature are commonly encountered, particularly as water is a by-product of cure.

(2) Vaporization of moisture occluded or otherwise contained within the cured or uncured polymer compositions is not easily controlled, and may produce a number of undesired side effects. For example, at temperatures on the order of 300°F., "interfacial sponging" may occur at a bonding interface causing the cured material to lift off the surface to which it is bonded. When a cured body is completely contained, exposure to elevated temperatures may cause undue expansion or undesired "forced thermal extrusion" of the material. .

Although the cured elastomers possess enhanced resistance to high temperatures, their electrical insulation properties become impaired at temperatures above about 185°F.

A process developed by G.D. Carpenter, G. Gregory, S.H. Kalfayan and I.P. Seegman (U.S. Patent 2,964,503; December 13, 1960; assigned to Products Research Company) gives a cure of polyalkylene polysulfide polymers at a much faster rate, such completely cured polymer possessing enhanced resistance to swelling and to loss of physical properties in the presence of water and at elevated temperatures, and correspondingly greater resistance to thermal extrusion and interfacial sponging. The method also gives cured elastomers from liquid polyalkylene polysulfide polymers that possess improved characteristics as electrical insulators.

Generally stated, the process is based on the concept that control over the rate of cure and a modification of the characteristics of the cured composition may be achieved by incorporating into a curing composition, composed essentially of a base mixture of polyalkylene polysulfide polymers and a soluble curing agent adapted to liberate anions of chromium when in solution, between about 2 and 50 parts of a modifying and solubilizing agent from the group of amides, sulfoxides, sulfones, sulfonamides, phosphoramides, esters of phosphoric acid, esters of boric acid, and esters of monobasic and polybasic organic acids per each 100 parts of polyalkylene polysulfide polymer. The modifying and solubilizing agent should be in liquid form, having a melting point below about 20°C. and a boiling point above about 125°C., and should be stable to oxidation in the presence of the curing component. More over, such agent should be at least a partial solvent for the curing salt.

The improved results obtained through this process can only result from use of solubilizing and modifying agents, typified by the data set forth in the table below, which serves to illustrate the process. In each of batches A, B, C, D, E and F, the same liquid polyalkylene polysulfide polymer was employed, such polymer having an average molecular weight of about 5,000 and a viscosity on the order of 35,000 to 45,000 centipoises. The compounding of each batch of the liquid polymer (which can be derived from dichlorodiethyl formal and trichloroethane or trichloropropane reacted with a metallic disulfide) was substantially the same, as indicated in the table, and the batches tested under identical, approved conditions. Batch A was used as a check and employs a curing system comprising sodium bichromate and water as an activator. Batches B, C, D, E and F similarly employ soluble chromates as the curing component but employ as activators various organic solubilizing and modifying agents.

	(A)	(B)	(C)	(D)	(E)	(F)
Formulation, parts by weight:						
Base Mix—						
Polyalkylene polysulfide polymer	100	100	100	100	100	100
Fillers (Titanium dioxide, Calcium carbonate, Calcium oxide, Ferric oxide	50	50	50	50	50	50
Wetting agent (Stearic acid, polyethylene glycol thioether)	1	1	1	1	1	1
Adhesive resin (Phenolic, epoxy resins)	5	5	5	5	5	5

(continued)

	(A)	(B)	(C)	(D)	(E)	(F)
Formulation, parts by weight:						
Curing Component—						
Sodium bichromate	7	--	--	--	--	--
Calcium bichromate	--	7	--	--	--	3.5
Magnesium bichromate	--	--	7	--	--	3.5
Stronitium bichromate	--	--	--	7	--	--
Cobalt bichromate	--	--	--	--	7	--
Solubilizing Agent—						
Water	7	--	--	--	--	--
Diethylformamide	--	7	--	--	--	--
Tributoxyethylphosphate	--	--	7	--	--	--
Dimethylsulfoxide	--	--	--	7	--	--
N-ethyltoluenesulfonamide	--	--	--	--	7	--
Dimethylacetamide	--	--	--	--	--	7
Test and Method:						
Minimum application life in hours	3	3	3	3	3	3
Ultimate cure time in days	30	14	14	14	14	14
Thermal softening as indicated by Shore hardness—						
Initial at 75°C.	55	57	50	50	53	55
At 325°F. for 5 min.	40	48	38	40	42	45
Interfacial sponging—						
At 225°F.	None	None	None	None	None	None
At 275°F.	Moderate	None	None	None	None	None
At 325°F.	Considerable	None	None	None	None	None
Swelling on emersion in water, percent (14 days at 140°F.)	30	3	5	2	4	4
Compression set, percent (24 hours at 10 psi and 120°F.)	4	3	6	4	6	3
Resistance to hydrocarbons, percent loss (7 days at 140°F. in 30% aromatics)	4.6	6.0	2.6	2.0	7.8	4.0
Electrical insulation characteristics (Volume resistivity in ohm-cm.)						
Initial at 75°F.	3×10^{10}	3×10^{10}	5×10^{12}	3×10^{9}	2×10^{11}	1×10^{10}
At 300°F.	4×10^{6}	4×10^{8}	6×10^{9}	1×10^{7}	2×10^{9}	2×10^{8}
At 75°F. after 14 days at 165°F. and 100% Relative humidity	$<1 \times 10^{5}$	4×10^{8}	5×10^{9}	1×10^{7}	9×10^{8}	2×10^{9}
Adhesion, pounds per inch	30	50	32	44	28	49

REACTIONS WITH POLYEPOXIDES

Polysulfide Polymers

Four processes that react polysulfide rubbers with polyepoxides came out of Thiokol, the first in 1957 and the subsequent ones about 10 years later. In the first of these, developed by E.M Fettes and J.A. Gannon (U.S. Patent 2,789,958; April 23, 1957; assigned to Thiokol Chemical Corporation) polythio polythiols and polythio polyhydroxy compounds are cured by reaction with polyepoxide compounds, i.e., compounds containing at least two epoxide groups. As a result satisfactory curing occurs and disadvantages are overcome.

The properties of the cured compounded products are outstanding in respect to their electrical properties. This is rather surprising because the electrical properties of the previously known polysulfide polymers are not outstanding. The products have the usual good solvent and chemical resistance of the polysulfide polymers as well as their low temperature properties.

These properties, as well as the ease of mixing and curing make the process useful in making protective coatings and envelopes which may be sprayed, painted or cast in place. The electrical uses are far greater than those open to the polysulfide polymers cured in the conventional manner. The use of these materials for potting compounds is clear in view of their ability to flow into place, completely covering and protecting every part in a transformer case. In addition, the use of a hard type of cured polymer permits the cured product to support and hold the parts in place. Finally, the mixture can be cold cured in the case. Another use is in cable blocks where electrical cables are kept under the pressure of an inert gas. In this case, the polymer can be mixed, injected into the cable and allowed to cure in situ to the desired properties. It is obvious that a hot cure inside a lead covered cable might prove difficult of achievement, so that the cold cure of the process offers a real advantage. This process is useful in general in producing metal- and solvent-free cured products. Another use is in the field of rocket propellants as a binder for the propellant.

The polythio polythiols may be symbolized by the formula

$$H(SRS)_x[SR'(SH)_nS]_yH$$

where x varies from 2 to 10, y varies from 0 to 10, n varies from 0 to 2. They are made in accordance with the teachings of U.S. Patent 2,466,963 to Patrick et al., dated April 12, 1949 (Chapter 1). It is to be realized that in any given quantity of the polythio polythiols or polythio polyhydroxy compounds, there is an almost infinite number of individual molecules and all of them do not have the same values of x and y, and that those values represent a statistical average. Very frequently, as shown by the specific examples hereinafter set forth, those average values are fractional rather than whole numbers. The radicals R and R' are selected from the group consisting of:

Curing

The radical R has a free valence of only 2 and R' a free valance equal to one of the integers 2, 3 and 4. The free valences are attached to different carbon atoms. Particular classes of the radicals are alkylene radicals, ethylenically unsaturated aliphatic hydrocarbon radicals, saturated aliphatic oxahydrocarbon and thiahydrocarbon radicals and araliphatic hydrocarbon radicals in which the free valences are in the aliphatic portion.

The polythio polyhydroxy compounds used may be symbolized by the formula:

$$HOR''S(SRS_x[SR'(SSR''OH)_nS]_ySR''OH$$

where x, y, n and R and R' have the definition previously given in connection with the symbol for the polythio polythiols, and R'' has the same generic definition as R but may be specifically different.

The polythio polyhydroxy compounds may be prepared from polythio polythiols reacting the latter with mercaptans having the general formula OHR''SH where R'' has the definition already given, in accordance with U.S. Patent 2,606,173 (see chapter 4). The two classes of compounds, namely, polythio polythiols and polythio polyhydroxy compounds are symbolized by the general formula:

$$F(SRS_x[SR'(SF)_nS]_yF$$

where n, x, y, R and R' have the definition already given and F is a radical of the group consisting of —SR''OH and S''SH.

The polythio polyhydroxy compounds have the distinct commercial advantage over the polythio polythiols of freedom from unpleasant odor and the process provides means of curing the polythio polyhydroxy compounds far superior to anything previously known, both as to method and products obtained. Many polyepoxide curing agents can be used.

The important or critical character of the curing agents is that they must have at least two epoxide groups. If they possess that critical property the remaining structure of the curing agents is unimportant. The proportion of the curing agent to the materials to be cured, that is, polythio polythiol or polythio polyhydroxy compounds varies over a wide range, depending to a large extent on the molecular weight of the initial polythio polythiol or polythio polyhydroxy compound.

The lower the molecular weight of the initial material the greater is the proportion of curing agent necessary to effect the desired curing action. In general, the proportion of polyepoxide curing agent varies from about 10 to about 50% of the total, but in some cases may be increased to as much as 70 or 80%, or decreased to 2 to 5%.

The curing reaction of this process does not require heating because it will occur at room temperature, e.g., 25°C. To accelerate the reaction, higher temperatures may be employed, although it is generally desirable not to use temperatures sufficiently high to cause vaporization or boiling of the curing agent. In general, temperatures of 25° to 150°C. are indicative of the range of temperatures of the curing reaction, the higher temperatures in this range being employed with polyepoxide curing agents having sufficiently high boiling points, or low vapor pressures. The curing reaction is facilitated by aliphatic amines and that class of compounds in general is used as catalysts.

However, the use of such catalysts is not critical because the curing action will proceed in the absence thereof.

Example: 100 parts of curing agent Shell Epon 1204-2 and 1.5 parts of diethylene triamine were mixed with 25 parts of polythio polythiol. This material was cured by allowing to stand 4 hours and 30 minutes at room temperature followed by 15 minutes at 70°C. and then 30 minutes at 100°C. The relatively large amount of curing agent was necessitated by the very low molecular weight of the polythio polythiol. At the end of the cure there was none of the characteristic odor of the polythio polythio and the product was a nontacky, hard tough polymer. This loss of odor indicated that the polythio polythiol was actually cured into the compound and was not merely an admixture with the curing agent.

If the above polythio polythiols were cured by the conventional method, that is, by the use of lead peroxide, there would be danger of combustion since a large proportion of peroxide would be necessary to effect the cure. A large proportion of the curing agent is also used in this example because of the very low molecular weight of the polythio polythiol. However, the cure proceeds smoothly without any danger of combustion.

Aminopolysulfide Polymers

A process developed by M.B. Berenbaum (U.S. Patent 3,322,851; May 30, 1967; assigned to Thiokol Chemical Corporation) concerns the preparation and use of amino-functional polysulfide polymers, and especially their reaction with polyepoxides at economical cure rates without amine-catalysts to provide useful resinous products with enhanced properties of flexibility, stability, and utility.

The amino-functional polysulfide polymers may be described as that class of compositions having molecular structures depicted by the formula $R^1-L-(RSS-)_x R-L-R^2$, wherein R^1 and R^2 are the specific alkylamino groups as follows:

$$\left(\begin{matrix} R^4 & R^3 & H \\ | & | & | \\ -CH-CH-N- \end{matrix}\right)_z \begin{matrix} R^4 & R^3 \\ | & | \\ CH-CH-NH_2 \end{matrix}$$

z being one of 0, 1, 2, 3 and 4, R^3 and R^4 are the same or different and are radicals chosen from the group consisting of hydrogen, normal and branched alkyl radicals, and substituted and unsubstituted aromatic and alicyclic radicals, x is a positive number greater than 1 and R is an intervening polyvalent organic radical.

Example 1:   Amino-functional polysulfide polymer, AFP-1, was prepared as the reaction product of a low molecular weight (about 500 to 700) liquid mercaptan terminated polysulfide polymer, designated LP-8 polysulfide polymer and having essentially the structure

$$HS(C_2H_4-O-CH_2-O-C_2H_4-SS)_4-C_2H_4-O-CH_2-O-C_2H_4-SH$$

and ethylenimine in molar charge ratio of 1:2 as follows: Under a blanket of nitrogen gas, a one-liter resin pot equipped with a mechanical stirrer, reflux condenser open to the atmosphere, addition funnel, and gas inlet and outlet tubes, the latter connected to a 0°C.

cold trap, was charged with 335 g. (0.5 mol) of LP-8 polysulfide polymer (mercaptan content 9.85% by weight) and 47.3 g. (1.1 mols, providing a 10% mol excess of imine) of ethylenimine. The mixture was stirred without heating for about 2 hours, and then at 45° to 50°C. for an additional 2.5 hours.

Unreacted volatiles were removed from the reaction mass by distillation and vacuum treatment. The polymer product obtained, AFP-1, was a brown viscous liquid having a mercaptain content of 3.62% and a nitrogen content of 5.47%. This corresponded to a polymeric product having an average molecular weight of about 756. Product weight yield was 95.2% of theoretical.

Example 2: Amino-functional polysulfide polymers AFP-2 and AFP-3 were prepared as the reaction products of a mercaptan terminated polysulfide polymer having a molecular weight of about 990, designated LP-3 polysulfide polymer and having essentially the structure

$$HS(C_2H_4-O-CH_2-O-C_2H_4-SS)_6-C_2H_4-O-CH_2-O-C_2H_4-SH$$

and ethylenimine as follows: Under a blanket of nitrogen gas, a one-liter resin pot, equipped as above, was charged with 495 g. (0.5 mol) of LP-3 of polysulfide polymer (mercaptan content 6.67% by weight) and 47.7 g. (1.11 mols) of ethylenimine. The mixture was stirred without heating for about 2.5 hours, the color changing from dark brown to salmon, and then at 40° to 55°C. for an additional 7.5 hours.

Upon further heating for two hours at distillation temperatures of 56° to 130°C. no distillate was collected. The slightly clouded liquid pot product was filtered and then treated with mild heating under vacuum to produce AFP-2, an amber viscous liquid polymer having a mercaptan content of 1.95% and a nitrogen content of 1.59%. Product weight yield 471.5 g. was approximately 87.7% of theoretical.

AFP-2 was further treated with ethylenimine to provide an amino-functional polymer, AFP-3, having no reactive mercaptan groups: 451 g. of AFP-2 and 23.3 g. (0.538 mol) of ethylenimine were reacted together following the above procedure described for the preparation of AFP-2. The amber viscous liquid polymer product AFP-3 was obtained in 440.8 g. yield, and had a mercaptan content of 0.0% and a nitrogen content of 2.51% by weight.

AFP-3 amino-functional polysulfide polymer and an epichlorohydrin-bisphenol A type polyepoxide of approximate epoxide equivalent of 185, designated Epon 828, were mixed in various proportions and permitted to cure. The solid resinous products obtained were compared with those obtained by an otherwise identical cure process using comparable amounts of the prior art cure system LP-3 mercaptan terminated polysulfide polymer-Epon 828 polyepoxide-tri(dimethylaminomethyl)phenol.

The recipes used and properties obtained are described in the table on the following page, where pbw is parts by weight.

RESINOUS CURE PRODUCTS OF AFP-3 AND LP-3 POLYSULFIDE POLYMERS WITH POLYEPOXIDE EPON 828

	1	2	3	4	5	6	7
Recipe:							
AFP-3, pbw	100	145	200	--	--	--	--
LP-3, pbw	--	--	--	100	145	200	100
Epon 828, pbw	--	100	100	100	100	200	100
Tri(dimethylamino-methyl)phenol, pbw	--	--	--	10	10	10	--
Cure Properties:							
80°F. Cure time, hrs.	48	48	48	0.5	1	2	(1)
Hardness, Shore A:							
After 2 days at 80°F.	42	58	54	95	96	57	(1)
After 2 hours at 212°F.	46	71	62	90	97	57	(1)

[1]No cure.

Azomethine

In a related process developed by E.R. Bertozzi (U.S. Patent 3,331,816; July 18, 1967; assigned to Thiokol Chemical Corporation) an amino-functional azomethine-containing polysulfide polymer was prepared. A 500 ml. three-necked round flask fitted with stirrer, thermometer, and condenser open to the atmosphere was sequentially charged with (a) 120 g. (about 0.12 mol) of a dimercaptan functional polysulfide liquid polymer of approximately 1,000 molecular weight and predominantly composed of the repeating unit $+C_2H_4OCH_2OC_2H_4SS+$ and containing approximately 2% by weight of cross-linking units effected by trichloropropane, and (b) 12 g. (about 0.214 mol) of the alkenal acrolein (propenal).

The reactants were stirred at ambient temperatures for 10 minutes, during which their temperature rose to about 40°C. and then fell. The reactant mixture was then heated to about 50°C. and maintained thereat for about 2 hours with stirring. The viscosity of the pot mixture increased somewhat during this interval, and the odor of acrolein was still perceptible at its end.

The liquid pot mixture was transferred to a dropping funnel, and admitted slowly with stirring to 30 g. (about 0.204 mol) of triethylenetetramine in a clean 500 ml. three-necked round bottom flask fitted as decribed above over a period of 1 hour 20 minutes. The temperature of the pot contents was maintained at 50°C. during this interval. The temperature was then permitted to fall to ambient, and stand thereat for two to three days. The pot product obtained was a ruddy colored rather viscous liquid polymer that smelled amine. The product was then heated slowly to about 90°C. under a reduced pressure of about 50 mm. Hg, from which about 1.2 ml. of water was distilled. Analysis yielded approximately 5.67% by weight of nitrogen, as compared with an approximate theoretical 7% nitrogen for an amino-functional azomethine containing polymer of molecular weight of about 1,200 to 1,300 of the formula:

$$\left[\left(SC_2H_4OCH_2OC_2H_4SS \right)_{x/2} - C_2H_4OCH_2OC_2H_4S - C_2H_4CH=N - \left(C_2H_4NH \right)_2 - C_2H_4NH_2 \right]_2$$

<u>Resinous Cure Products with Polyepoxide:</u>   The cure properties and products of the liquid amino-functional azomethine containing polysulfide polymer produced above, and hereinafter called Polymer A, and a liquid bisphenol-A-epichlorohydrin polyepoxide with epoxide equivalent of 180 to 200 and viscosity at 77°F. of 90 to 180 poises, and hereinafter called Epi-Rez 510 Polyepoxide, were observed when they were admixed under the conditions and in the proportions listed in the following table.

Experiment	(a)	(b)
Recipe in parts by weight:		
Polymer A	100	100
Epi-Rez 510 Polyepoxide	200	100
Properties at 80° F.:		
Liquid Pot Life, minutes	110	80
Set or Gel Time, minutes	125	90
Tack-Free Time of Cured Products, hours	8	3 to 5
Highest Temperature during Cure, ° F	107	127
Odor of Recipe	(1)	(1)
Cured Casting:		
Color	Orange	Orange
Clarity	(2)	Clear
Hardness, Shore "D" Durometer Degrees at 80° F.:		
After 72 hours	33	65
After 7 days	51	72
After 14 days	83	77
At 120° F., measured at 80° F.:		
For 72 hours	81	77
For 14 days	83	78

1 Slightly Amine.
2 Slightly Hazy.

From the foregoing, one may note that the preferred process for providing the amino-functional azomethine containing polysulfide polymers entails the steps of admixing the chalcogen-hydric functional polysulfide polymer and the alkenal, heating and/or permitting the temperature of the mixture produced to at least about 40°C. for about 2 hours, admixing the subproduct formed with the poly-primary amine, and heating this admixture to at least about 50°C. for at least about 1 hour.

Amino-Functional Phenol

Still another similar process developed by E.R. Bertozzi (U.S. Patent 3,335,201; August 8, 1967; assigned to Thiokol Chemical Corporation) gives an amino-functional phenol-containing polysulfide polymer prepared as follows.

A 500 ml. three-necked round bottom flask fitted with stirrer, thermometer, and distillation condenser open to the atmosphere, was sequentially charged with (a) 120 g. (about 0.12 mol) of a dimercaptan functional polysulfide liquid polymer of approximately 1,000 molecular weight and predominantly composed of the repeating unit

$$+C_2H_4OCH_2OC_2H_4SS+$$

and containing approximately 2% by weight of cross-linking units effected by trichloropropane, (b) 20 g. (about 0.213 mol) of phenol, (c) 14 g. (about 0.465 mol) of formaldehyde in the form of its trimer paraformaldehyde, and (d) 30 g. (about 0.204 mol) of triethylenetetramine, an amine having two primary and two secondary amine groups each separated by an ethylene linkage $+C_2H_4+$. The reactants were heated with stirring at 100°C. for about 2 hours. The temperature of the reaction mixture was permitted to fall to ambient and remain there for two to three days. The reaction mixture was then heated to 100°C. under reduced pressure of about 100 mm. Hg, at which point about 3 ml. of water were distilled. The pot product remaining was an orange colored viscous liquid polymeric material with a faint phenolic and ammoniacal odor. Analysis of the polymeric product for weight percent nitrogen was determined to be about 5.7%, as compared to about 6.3% theoretical for an amino-functional phenol containing polymer of formula

$$-CH_2 \overset{OH}{\underset{}{\bigcirc}} CH_2-NH(C_2H_4NH)_x-C_2H_4NH_2$$

$$-[S-(C_2H_4OCH_2OC_2H_4SS)_nC_2H_4OCH_2OC_2H_4S]-CH_2 \overset{OH}{\underset{}{\bigcirc}} CH_2-NH(C_2H_4NH)_xC_2H_4NH_2$$

and of molecular weight of about 1,200 to 1,300.

Resinous Cure Products with Polyepoxide: The cure properties and products of the liquid amino-functional phenol containing polysulfide polymer produced above, and hereinafter called Polymer A, and a liquid bisphenol-A-epichlorohydrin polyepoxide with epoxide equivalent of 180 to 200 and viscosity at 77°F. of 90 to 180 poises, and hereinafter called Epi-Rez 510 polyepoxide, were observed when they were admixed under the conditions and in the proportions listed in the following table.

Experiment	A	B
Recipe, in parts by weight:		
Polymer A	100	100
Epi-Rez 510 Polyepoxide	200	100
Properties at 80° F.:		
Liquid Pot Life, minutes	120	70
Set or Gel Time, minutes	130	75
Tack-Free Time of Cured Product, hrs	24	4
Highest Temperature during Cure, ° F	96	118
Odor of Recipe	(1)	(1)
Cured Casting:		
Color	Amber	Amber
Clarity	Clear	Clear
Hardness, Shore "D" Durometer Degrees at 80° F.:		
After 72 hours	22	56
After 7 days	53	67
After 14 days	86	76
At 120° F.:		
For 72 hours	84	76
14 days	84	76

[1] Slightly Amine.

Poly-primary amines are used which contain at least two primary amine groups separated by alkylene groups having of up to about 6 carbons, alkylene-secondary amine-alkylene groups having up to six carbon atoms in each alkylene linkage, and benzyl and alkylenebenzyl groups having up to about six carbon atoms in the alkylene portion. The preferred poly-primary amines are those having two primary amine groups that may be separated by alkylene or alkylene-secondary amine-alkylene groups as in the formula

$$H_2N(R_c-NH)_d R_c-NH_2$$

wherein R_c is an alkylene group having up to six carbon atoms and d is an integer that is one of 0, 1, 2, 3, 4 and 5. The most preferred useful poly-primary amines are those depicted by the foregoing formula wherein R_c is an ethylene linkage and d is one of 1 and 2. The presence of the secondary amine groups in the preferred amines aids in yet further accelerating the reaction of the polysulfide polymers formed with polyepoxides.

MANGANITE-COATED MnO_2

In a process developed by N.A. Rosenthal, J.R. Panek and K.R. Cranker (U.S. Patent 2,940,959; June 14, 1960; assigned to Thiokol Chemical Corporation) polysulfide polymers, especially polymers having low molecular weights of the order of about 1,200, can be effectively and controllably cured to form solid, rubber-like polysulfide resins of superior temperature stability by intimately admixing with such polymers a manganese dioxide activated for that purpose by the formation of an alkali manganite on the surface of the manganese dioxide. An activated manganese dioxide capable of accelerating and effecting a cure of polysulfide polymers of molecular weights as low as about 1,000 and as high as 100,000 and possibly 500,000 or 600,000 can readily be prepared by precipitating manganese dioxide from an alkali metal permanganate solution under alkaline conditions.

Examination of the manganese dioxide obtained in accordance with this procedure indicates that it is composed of minute particles of manganese dioxide having a relatively large surface area coated with an alkali metal manganite. Tests have revealed that control of curing rates with the activated manganese dioxide of the process can readily be effected, the rate being retarded in the presence of materials such as stearic acid, ammonium salts, nitro compounds, alums, etc. and accelerated in the presence of amines, alkali metal oxides and hydroxides, and the like.

To illustrate the best mode of preparing manganese dioxide activated for the purpose of effecting a polysulfide polymer cure, 632 grams $KMnO_4$ and 1,600 ml. of water in a five-liter flask were heated to 60°C. The flask was then removed from the heat source and a solution containing 250 grams Na_2SO_3 in one liter of water was added. During this addition, the temperature rose to 96°C., at which time a solution containing 150 grams NaOH in 500 ml. water was added. The purple solution following this treatment became brownish purple. 500 grams Na_2SO_3 were then slowly dusted into the solution while stirring, the rate of addition being adjusted so that the temperature remained below 106°C. A solution of 150 grams NaOH in 200 ml. water was then added, the pH of the resulting solution being 11.0. 265 grams of $KMnO_4$ were slowly added to the reaction mixture followed by addition of a solution of 100 grams NaOH in 250 ml. water.

400 grams Na_2SO_3 were then added slowly while the temperature rose to 102°C. and the color changed to a greenish brown. Finally, an additional 600 grams Na_2SO_3 were slowly added, at which point the color of the solution changed to a deep brown without a change in temperature. The solution was then cooled to room temperature and filtered with a Buchner funnel, the total amounts of reactants used being 400 g. NaOH, 897 g. $KMnO_4$, 1,750 g. Na_2SO_3, and 3,550 g. H_2O. An MnO_2 precipitate was then removed from the filter and washed with distilled water until the pH of the wash liquid was reduced to about 8, dried, washed with acetone and finally air-dried to yield 680 grams of activated

manganese dioxide. In effecting a cure of polysulfide polymer, the activated MnO_2, and other materials such as fillers, accelerators or retarders and the like are intimately admixed in any suitable mixing device such as a paint mill and then allowed to cure at room or elevated temperatures. The proportions of MnO_2 utilized in accordance with this method are not critical and may vary considerably, depending upon the particular polymer and curing conditions. Generally, however, about 4 to 10% are considered optimum.

CURE FOR LIGHT COLOR

Use of Antimony Trioxide

A process developed by W.E. Leuchten and A.F. Vondy (U.S. Patent 3,036,049; May 22, 1962; assigned to Thiokol Chemical Corporation) is particularly concerned with liquid polysulfide polymers which can be obtained from the high molecular weight polymers by a splitting process described in U.S. Patent 2,466,963 (see Chapter 1). As disclosed in the latter patent, the high molecular weight polymers can be split to form polythiopolymercaptan polymers having molecular weights of the order of 500 to 25,000 and which are viscous liquids having viscosities within the range 300 to 100,000 centipoises. Such liquid polymers can be cured by any of various curing agents as disclosed in U.S. Patent 2,466,963.

Such liquid polymers when mixed with a curing agent such as, for example, lead peroxide, can be applied to a metal surface and cured at room temperature or at an elevated temperature to form a strongly adherent rubber coating on the metal. However, it has not been possible to produce a satisfactory coating of this type which is either white or pastel-colored. When the previously proposed curing agents and curing procedures are used, a discolored coating or one that is not sufficiently adherent to the metal surface is obtained. In certain of the sealant applications wherein these materials are used it is highly desirable that the cured coating be white or colored some light pastel color. It has not previously been possible to achieve this result.

A cure to obtain light colors can be achieved by using antimony trioxide as a curing agent for a liquid polysulfide polymer of the type described above. It has been found that when antimony trioxide is used for curing such polymers, a very light colored elastomer is obtained that is strongly adherent to metal surfaces. This result is quite surprising since antimony pentoxide which contains a higher proportion of oxygen than the trioxide does not give a satisfactory cure.

The antimony trioxide used may be either the pure compound or any of various commercial products which contain small amounts of impurities such as zinc oxide and tin oxide. The presence of such minor amounts of impurities does not appear to interfere with the curing of the polymer to achieve the desired light-colored product. In order to achieve a practical cure rate, it is necessary that the particle size be sufficiently small. Commercially available Sb_2O_3 may be satisfactory but it is generally desirable to ball mill the material for 24 hours before using. Excess ball milling or ball milling an already satisfactory lot is harmless. In general the curing procedure employed is similar to that previously used for curing liquid polysulfide polymers with curing agents such as lead peroxide.

Curing

Example 1:   The polysulfide polymer used in this example is a liquid polymer having a molecular weight of about 4,000. It may be prepared by following a procedure in U.S. Patent 2,466,963. To 100 parts of this polymer dispersed on a paint mill there were added in sequence 35 parts by weight of titanium oxide pigment (Titanox AMO), 10 parts by weight of finely divided silica (Cab-O-Sil), 15 parts by weight of substantially pure antimony trioxide (Sb_2O_3) and 5 parts of sodium stearate as a buffer.

Mixing was continued for one-half hour to insure thorough dispersion of the solid materials of the liquid polymer. The mixture was allowed to stand for six hours to effect a partial cure and then pressed out into test sheets 0.075 to 0.080 inches thick. The test sheets were cured for an additional period of 16 hours at room temperature, at the end of which time curing of the polymer to an elastomer appeared to be complete. The average tensile strength of the test sheets was 370 psi and their maximum elongation was 550%. The test pieces exhibited a hardness of 43 on the Shore A scale.

Example 2:   The polymer used in this example had a molecular weight similar to that of the polymer of Example 1, but differed from the Example 1 polymer in that it had 2% of cross-linking rather than 0.5% of cross-linking. The polymer of this example is sold under the commercial designation LP-2.

This polymer can be used with advantage in the making of dental impressions. However, the usual lead peroxide cure is objectionable in this application because of the relatively dark brown color of the resulting elastomer. By employing the following formulation containing antimony trioxide as a curing agent a light-colored and therefore more acceptable molded product can be achieved.

	Parts by Weight
Liquid polymer	100
Titanox AMO	50
Sb_2O_3	40
ZnO_2	10
PbO	10

This composition sets in 5 minutes and cures in 8 minutes to give a pale yellow rubbery material.

Example 3:   Pastel shades may be obtained by using small amounts of paste pigments of the desired color. The use of such pigments is illustrated by the formulations below wherein the polymer is the same as that of Example 1 and the quantities are given in parts by weight.

Liquid Polymer	100	100	100
Titanox AMO	35	35	35
Cab-O-Sil	10	10	10
Sodium Stearate	5	5	5
Sb_2O_3	15	15	15
Claremont Paste:			
4050-CSL-1 (Pink)	0.4		
4040-UB-2 (Blue)		0.4	
4060-HCG-1 (Green)			0.4
Color of Cured Product	Pink	Blue	Green

Use of Dibutyl Tin

Another process developed by A.F. Vondy and W.E. Leuchten (U.S. Patent 3,243,403; March 29, 1966; assigned to Thiokol Chemical Corporation) uses dibutyl tin oxide as a curing agent for a liquid polysulfide polymer. It has been found that when dibutyl tin oxide is used as a curing agent, compositions are obtained wherein the cure rate is accelerated under acid conditions and retarded under alkaline conditions, in contrast to previous cure formulations wherein the cure rate is retarded under acid conditions and accelerated under alkaline conditions. Also, when dibutyl tin oxide is used as a curing agent, the usual phenolic additive adhesion-promoters can be incorporated in the formulation to yield a composition which not only cures to a light-colored elastomer but also gives a good adhesion to substrate surfaces without the use of a preliminary primer coat.

Example 1: The polysulfide polymer used in this version was a liquid polythiopolymercaptan polymer having a molecular weight of about 4,000 and about 2% cross-linking. It may be prepared in the manner described in the introductory portion of the specification and is sold under the commercial designation LP-2. 100 parts of this liquid polymer were mixed with 25 parts of white powdered dibutyl tin oxide (Thermolite-15 manufactured by Metal and Thermit Corporation). This mixture was placed in a 70°C. oven for two hours and produced a white cured elastomer of a tough, resilient and flexible nature.

Example 2: The polysulfide polymer used in this example was similar to that used in Example 1 except that it had only about 0.5% of cross-linking. It is a commercially available polymer sold under the trade designation LP-32. Five formulations containing this polymer were prepared having the compositions indicated in the table below. In each case 100 parts by weight of the liquid polymer were mixed with 50 parts by weight of pigment and different amounts of dibutyl tin oxide as indicated in the table. The filler was a titanium oxide powder having a particle size of 0.1 to 0.3 micron sold under the trade name Titanox AMO. The dibutyl tin oxide was used in the form of a paste composed of 2 parts by weight of dibutyl tin oxide and 1 part by weight of dibutyl phthalate.

The five formulations were cured at a temperature of 75°F. ± 2° and a relative humidity of 50%. The working life, set time and cure time of each sample was determined and the hardness of each cured sample was measured with a Shore Durometer using the "A" scale. The results are summarized in the table wherein all quantities are given in parts by weight. The working life, set time and cure time are expressed in hours.

Component of Formulation	1	2	3	4	5
LP-32 liquid polymer	100	100	100	100	100
Titanox AMO	50	50	50	50	50
Dibutyl Tin Oxide paste	7.5	11.25	15	22.5	30
Amount of Dibutyl Tin Oxide in paste	5	7.5	10	15	20
Properties:					
Working Life	1	0.83	0.42	0.5	0.5
Tack-free set time	6	2.5	2	1	0.75
Cure Time	16	6	4.5	3	3
Shore "A" Hardness, after 24 hours	8	20	35	35	35

The data of the foregoing table show the manner in which variations in the amount of dibutyl tin oxide produced changes in the working life, set time and cure time of the different formulations. It may be noted that Shore "A" hardness of 20 is considered a soft cure and a hardness of 35 is considered a reasonably firm cure.

Example 3: Formulations of liquid polythiopolymercaptan polymers incorporating dibutyl tin oxide as a curing agent are useful in the coating of surfaces to provide elastomeric protective films. The coatings may be applied in the usual manner by spraying, brushing or dipping techniques. Dibutyl tin oxide as a curing agent makes possible formulations which both give a "white" cure and are compatible with phenolic resin adhesion-improving agents.

This modification illustrates formulations of this type. The phenolic additive incorporated in the formulations of this example was a phenolic resin identified by the trade name Durez 10694. The formulations given below also include the solvents methyl ethyl ketone and toluene to give the spray compositions the desired fluidity and varying amounts of stearic acid to illustrate the manner in which the cure rates are accelerated under acid conditions in the formulations. Cure times are given in minutes.

Components of Formula	1	2	3
LP-32	100	100	100
Titanox AMO	50	50	50
Durez 10694	5	5	5
Stearic acid	0.5	1.0	1.5
MEK	30	30	30
Toluene	30	30	30
Dibutyl Tin Oxide	10	10	10
Dibutyl Phthalate	5	5	5
Properties:			
Working life	15	10	8
Tack-free set time	29	20	18
Cure time	145	125	115

The data clearly bring out the fact that as the amount of stearic acid is increased the working life, tack-free set time and cure time of the mixtures all decrease. They demonstrate the accelerating effect of a low pH material on the cure rates of formulations employing dibutyl tin oxide.

Example 4: This example illustrates formulations suitable for use in preparing dental impressions, wherein a "white cure" and exceptionally high cure rates are desirable. The compositions and properties of six formulations of this type are given. Formulations 1 and 2 contain a calcium carbonate pigment which is essentially ground oyster shells and is sold under the trade name "Laminar."

Formulations 3 and 4 contain a wet ground calcium carbonate pigment sold under the trade name "Camel Wite." Both of these calcium carbonates have a pH of approximately 8.5. Formulations 5 and 6 of the table on the following page contain a titanium dioxide pigment having a pH of about 7. The polysulfide polymer used in these formulations is the same as that of Example 1.

Components of Formula	1	2	3	4	5	6
LP-2	100	100	100	100	100	100
Laminar	10	20				
Camel Wite			10	20		
Titanox AMO					10	20
Dibutyl Tin Oxide	140	140	140	140	140	140
Dibutyl Phthalate	70	70	70	70	70	70
Properties:						
Working Life in minutes	2	3	2	3	3	3
Set Time in minutes	3	4	3	4	4	4
Tack-free Time in minutes	4	6	5	6	5	5
Cure Time in minutes	5	8	7	8	6	6

For dental impressions it is desirable that formulations cure over a period of the order of five to ten minutes, and the data show that the formulations cure in periods of this general magnitude. The white pigments used produce pleasant appearing white formulations which cure to resilient compounds having a hardness as measured by the Shore "A" durometer of approximately 30.

USE OF SULFONE-ACTIVATED DIETHYLENIC COMPOUND

A process developed by G.M. Le Fave and F.Y. Hayashi (U.S. Patent 3,138,573; June 23, 1964; assigned to Coast Pro-Seal & Mfg. Company) uses a sulfone activated diethylenic compound as an agent for producing a vulcanized or cured elastomer from liquid, high molecular weight polythiol compounds. In the process, the polythiol is cast in liquid state and is cured in place by the sulfone to a final elastomeric state, with or without the application of heat.

The sulfone activated diethylenic compound, which is the acceptor in the reaction, is preferably one which has ethylenic ($-C=C-$) linkages activated by a sulfone ($-SO_2-$) group adjacent the ethylenic linkages. The preferred specific acceptor is divinyl sulfone.

Example 1: To 193 grams polysulfide liquid polymer, equivalent weight 322, (0.6 equivalent) was added 0.23 gram (0.1%) triethylene diamine. After warming the mixture slightly so that the amine would be readily dissolved, the mixture was brought to room temperature. 35.4 grams (0.6 equivalent) divinyl sulfone were added and stirred in. After an exotherm to about 90°C., the liquid was poured into molds and allowed to cure at ambient temperature.

	1 Week	4 Weeks
Hardness (Shore "A")	29	46
Tensile (psi)	67	120
Elongation (ultimate), percent	170	100

Example 2: To illustrate the advantages of having a trithiol compound, rather than a dithiol functionality, and the effect that this functionality has upon the amount of divinyl sulfone needed to bring about an equivalent cure, the following comparison was made:

Two polysulfide liquid polymers were each reacted with varying amounts of divinyl sulfone and allowed to cure. These polysulfide resins each have an average molecular weight of

1,000 and differ only in their degree of trifunctionality or cross-linking. Polymer No. 1 was made from 98 mol percent of bis(2-chloroethyl) formal and 2 mol percent of trichloro-propane, the cross-linking agent, while Polymer No. 2 was prepared from 99.5 mol percent of the same formal and 0.5% of trichloropropane. Hence, it can readily be seen that the degree of cross-linking is much greater in the case of Polymer No. 1. In the following examples, 0.1% triethylene diamine was used in each instance, and the procedure used was as in Example 1.

Property, 1 Week	Liquid Polymer	Ratio of Equivalents, D.V.S.: Liquid Polymer			
		0.0971:1	1.0:1	1.025:1	1.06:1
Hardness(Shore"A")	No. 2	*	*	20	25
	No. 1	10	18	27	40
Tensile (p.s.i.)	No. 2	*	*	53.2	74.9
	No. 1	15.0	28.4	48.0	97.8
Elongation (percent)	No. 2	*	*	290	260
	No. 1	190	190	85	85

*Not sufficiently cured to take reading.

USE OF ORTHONITROANISOLE

Orthonitroanisole and Metal Oxides

It is known that the use of nitroaryl compounds with metal oxide curing agents provides a synergistic effect in the vulcanization of liquid polysulfide polymers. The amounts of each of the metal oxide and nitroaryl compounds needed to produce the synergistic effect when they are used in combination are, in general, substantially less than the amounts of either that are needed when they are used alone as vulcanizing agents. The use of metal oxide/nitroaryl compound cure systems, however, has certain disadvantages. For example, such combined curing systems will, upon admixture with the liquid polysulfide polymers, rapidly cure the polymers at room temperature, e.g., 25°C.

The curable compositions will thus thicken from a mixture of workable consistency, to a heavy syrup or to a gelled or set condition of unworkable consistency at room temperature in a very short time. The attainment of the unworkable consistency usually occurs over a period as short as one half to two hours. It is only during the interval of relative workability of the cure compositions and prior to gellation that the polymer systems may be readily worked to coat or pot articles, or to fill molds. The progressive and rapid thickening of the prior art curable polysulfide compositions during the working life also makes divestiture of occluded and dissolved gas therefrom most difficult if not impossible, and thus to provide in the end rubber articles having undesirable surface pock marks and internal voids.

According to a process developed by J.R. Panek (U.S. Patent 3,282,902; November 1, 1966; assigned to Thiokol Chemical Corporation), if orthonitroanisole is used to the extent of about 2 to 50 parts by weight per 100 parts by weight of liquid polysulfide polymers of the type revealed in U.S. Patent 2,466,963 (see again Chapter 1) in admixture with those metal oxide curing agents known to be useful in promoting the vulcanization of liquid poly-sulfide polymers that (1) curable compositions are formed which have prolonged working lives at room temperatures, e.g., about 80°F., which often extend into days; (2) the cur-able compositions will quickly cure to form elastomers with good physical properties at elevated temperatures of 140° to 200°F., in 1/2 to 8 hours, the longer intervals occurring at the lower curing temperatures; (3) upon cure, vulcanizates are produced which are ex-traordinarily bloom resistant for extended periods of time.

Other important benefits were found to obtain through the use of orthonitroanisole as the nitroaryl compound component of the curing agent system. Other nitroaryl curing agent materials require a minimum of three passes on a paint mill for uniform incorporation into the curable compositions; orthonitroanisole requires but one. Further, the rapidly thicken-ing curable compositions containing the nitroaryl compounds of the prior art makes it dif-ficult to divest them of the dissolved and occluded gases, whereas, with the compositions of this process, removal of gases is made relatively simple and rapid because of their pro-longed work life.

Examples 1 to 7: In the following examples, an intimate blend of the ingredients of the curable compositions was prepared by milling together the polysulfide liquid polymer with the nitroaryl compound used and at least a portion of the other adjuvants as one partial mix-ture, and by milling together the metal oxide in a liquid vehicle with perhaps some of the other adjuvants as the second partial mixture, then combining the two partial mixtures by hand mixing at room temperature to form the specific curable composition. The working life of the curable compositions was observed to be that interval at room temperature from the time of mixing to the time when the compositions were no longer pourable.

The cure time was observed to be that interval at a specified elevated temperature in which one obtained a tack-free elastomeric solid. The cure times listed herein are merely times which happened to be observed and are not necessarily the minimum cure times required. In general, the compositions were cured to form vulcanizates in the form of test sheets. These were examined for physical property data, such as tensile properties (ASTM D412-51T) and hardness (ASTM D676-59T).

Polysulfide polymers of different molecular weights were used; LP-8 polysulfide liquid poly-mer has a molecular weight of about 600, a 2% cross-link with trichloropropane and a back-bone that is essentially

$$HS(C_2H_4OCH_2OC_2H_4SS)_nC_2H_4OCH_2OC_2H_4SH$$

LP-3 and LP-2 polymers are similar to LP-8 polymer, except that the molecular weights are higher, i.e., about 1,000 for LP-3 polymer and about 4,000 for LP-2 polymer. The nitro-aryl compounds used were commercial grade chemicals. The manganese dioxide used in these examples was Manganese Hydrate No. 37, a substance being 70% MnO_2, 22% water of hydration, 43.8% manganese and 12.5% in available oxygen. The tellurium dioxide used was analyzed as 95% TeO_2. The solvent vehicles used were Aroclor 1254, a poly-chlorinated biphenyl of 54% chlorine, and HB-40, a partially hydrogenated terphenyl. Occasionally a surface active agent, Duponol L-144-WDG was used. This is a sodium salt of a modified unsaturated long chain alcohol sulfate. All ingredients are given in parts by weight (pbw). Aroclor 1242 is chemically similar to Aroclor 1254, but is a less viscous

Example	1	2	3	4	5	6	7
Curable Composition Recipe, in pbw:							
LP-8 polysulfide polymer	100	—	—	—	—	—	—
LP-3 polysulfide polymer	—	—	100	100	100	100	100
LP-2 polysulfide polymer	—	100	—	—	—	—	—
MnO$_2$	6	2.38	2.12	2.12	2.5	4.5	4.5
TeO$_2$	—	—	—	—	12	5	5
Orthonitroanisole	4	3	5	5	—	—	—
2,4-dinitrobenzene	—	—	—	—	—	—	—
Sterling MT, carbon black	—	0.6	30.5	30.5	30	30	30
Titanox RA-50, titanium dioxide	—	25	—	—	—	—	—
HB 40 terphenyl	—	1.8	1.6	1.6	2.5	—	3
Aroclor 1254 biphenyl	6	—	—	—	—	3	—
CuCl$_2$	—	0.01	0.01	0.01	—	0.17	0.17
Duponol L-144-WDG	—	—	—	—	—	—	—
Water	—	0.01	0.01	0.01	—	3	—
Aroclor 1242 biphenyl	—	—	—	—	—	—	0.17
Curing Conditions:							
Working life, at 80°F., in hours	>280	10	>25	0.8	>48	28.5	>1
Curing temperature, in °F.	180	158	180	180	180	180	—
Curing time observed, in hours	29	0.9	5	5	<24	2.25	—
Physical Properties of Vulcanizate:							
Conditioning temperature after cure, in °F.	—	—	—	—	180	180	180
Conditioning time after cure, in hours	—	—	—	—	24	16	16
Tensile strength, in psi	—	—	175	230	190	190	200
Elongation, in percent	—	—	260	270	370	265	270
Hardness, in Shore A durometer degrees	5	25	36	44	32	38	40

liquid and has a chlorine content of 42%. Examples 1 and 2 indicate that regardless of molecular weight, polysulfide liquid polymers will form curable compositions with orthonitroanisole having a long termed working life. Examples 3 and 4 demonstrate the differences in working life obtained with curable polysulfide polymer based compositions using orthonitroanisole and manganese dioxide on the one hand and a nitroaryl of the prior art, dinitrobenzene, with manganese dioxide on the other hand. Example 5 shows a curable polysulfide polymer based composition using orthonitroanisole near the upper end of the preferred range of concentration.

Examples 6 and 7 contrast the properties of curable polysulfide polymer based compositions, as a working life, using another curing agent, TeO_2, with orthonitroanisole in one, and dinitrobenzene in the other composition. In instances where orthonitroanisole was used, the vulcanizates were substantially bloom-resistant; whereas when dinitrobenzene was used the vulcanizates often exhibited bloom.

Orthonitroanisole, Metal Oxides, and Copper-Based Stabilizer

Despite the advantages of the process described in U.S. Patent 3,282,902 (above), difficulties are encountered with metal oxide/orthonitroanisole curing systems. These difficulties arise from an unusual sensitivity of curable polysulfide/metal oxide/orthonitroanisole systems to even small variations in quantity or type of the component ingredients used in a system recipe. This sensitivity is often manifest by inconsistent trends, both in the cure properties e.g., cure rate, and in the physical properties of the resulting vulcanizates with even small changes made in the amounts and types of recipe ingredients, such as that of the metal oxide curing agents and/or the orthonitroanisole curing aid, and/or the adjuvant materials.

It is also often manifest by a poor reproducibility of cure properties and of the physical properties of the resulting vulcanizates where apparently identical curing recipes and conditions are used, albeit at different times of day or week, or with different batches or lots of ingredients used. It is even manifest to the degree of occasionally providing no cure at all. The process concerns the remedy of such difficulties, and concerns in particular cure systems employing orthonitroanisole as the curing aid.

In general, the instabilities and inconsistencies exhibited in curing characteristics of polysulfide liquid polymers using orthonitroanisole as the curing aid are substantially assuaged through the use of a process developed by J.J. Giordano (U.S. Patent 3,349,057; October 24, 1967; assigned to Thiokol Chemical Corporation). This process involves a curing stabilizer system comprising per 100 parts by weight of polysulfide liquid polymer, 2 to 50 parts by weight of orthonitroanisole, at least 0.005 part by weight of copper derived from at least one polar-liquid-soluble copper salt and at least 0.01 part by weight of at least one polar liquid which liquid has a group dipole moment of at least 0.5 debye units. The stabilizer systems may be in the form (a) of solutions of copper salts in the polar liquids; (b) of finely dispersed suspensions of the copper salts partially dissolved in the polar liquids; and (c) of each of the aforesaid copper salts and the polar liquids used as separate recipe ingredients mixed together at the time of, and with the mixing of the composition ingredients.

The copper salts which may be used are those which are soluble in polar liquids to the extent of at least 5% by weight. This is what is meant by a polar-liquid-soluble copper salt. Where the solubility is less than 5% it is believed that not enough copper is made available to produce satisfactory stabilization of the curable compositions.

Examples 1 to 16: With the same conditions for blending ingredients as in U.S. Patent 3,282,902, Giordano's process produced the compositions shown in the following tables.

Example	1	2	3
Recipe Ingredients in pbw:			
LP-3 polysulfide polymer	--	100	100
LP-2 polysulfide polymer	100	--	--
Manganese dioxide	2.38	2.12	2.12
Orthonitroanisole	3	5	--
2,4-dinitrobenzene	--	--	5
Medium Thermal Furnace carbon black (Sterling MT)	0.6	30.5	30.5
Titanium dioxide (Titanox RA-50)	25	--	--
HB-40 terphenyl	1.8	1.6	1.6
$CuCl_2 \cdot 2H_2O$	0.01	0.01	0.01
Curing Conditions:			
Working life, at 80°F., in hours	10	>24	0.8
Curing temperature, in °F.	158	180	180
Curing time observed, in hours	0.9	5	5
Physical Properties of Vulcanizates:			
Ultimate tensile strength, in psi	(1)	175	230
Ultimate Elongation, in percent	(1)	260	270
Hardness in Shore "A" durometer degrees	45	36	44

[1] Not tested.

Examples 1 and 2 indicate that regardless of molecular weight, polysulfide liquid polymers will form curable compositions with orthonitroanisole and copper salt/polar liquid stabilizer systems which have desirable long termed working life, Examples 2 and 3 demonstrate the differences in working life obtained with curable polysulfide polymer based compositions using orthonitroanisole on the one hand, a nitroaryl of earlier methods, dinitrobenzene on the other hand. In instances where orthonitroanisole was used, the vulcanizates were substantially bloom-resistant; whereas when dinitrobenzene was used in vulcanizates often exhibited bloom. Beyond the working life periods indicated, at 80°F., the compositions would thicken and vulcanize to solid elastomers in less than 100 hours.

In Examples 4 to 16, on the following page, an intimate blend of the ingredients of the curable compositions was prepared by milling together the polysulfide liquid polymer with orthonitroanisole, and at least a portion of the other adjuvants as one partial mixture, and by milling together the manganese dioxide curing agent, copper salts, and polar liquids where used in a liquid vehicle with some of the other adjuvants as the second partial mixtures at room temperature to form the specific curable composition. Different curing pastes employing partial stabilizing systems of this process were used.

Example	4	5 (2)	6	7	8	9 (2)	10	11	12 (2)	13 (2)	14 (2)	15 (2)	16 (2)
Recipe in pbw:													
LP-3 polysulfide polymer	100	100	100	100	100	100	100	100	100	100	100	100	100
Sterling MT, carbon black	30	30	30	30	30	30	30.4	30.6	30.4	30.7	30.7	31.1	31.1
o-Nitroanisole	4	12	5	4	5	12	5	5	12	12	12	5	5
Manganese dioxide	1.5	1.5	1.5	2.5	2.5	3.0	1.43	2.38	1.43	2.86	2.9	4.28	5.7
Aroclor 1254 biphenyl	1.5	1.5	1.5	2.5	2.5	3.0	0.0	0.0	0.0	0.0	0.0	0.0	0.0
HB-40, terphenyl	0.0	0.0	0.0	0.0	0.0	0.0	1.12	1.88	1.12	2.24	2.12	3.38	4.5
$CuCl_2 \cdot 2H_2O$	0.0	0.0	0.0	0.0	0.0	0.0	0.07	0.12	0.07	0.14	0.0	0.24	0.29
$Cu(C_2H_3O_2)_2 \cdot H_2O$	0.0	0.0	0.0	0.0	0.0	0.0	0.0	0.0	0.0	0.0	0.07	0.0	0.0
Copper content of recipe	0.0	0.0	0.0	0.0	0.0	0.0	0.03	0.0445	0.03	0.06	0.02	0.08	0.108
Water added	0.0	0.0	0.0	0.0	0.0	0.0	0.07	0.12	0.07	0.14	0.26	4.24	0.29
Curing Conditions:													
Curing time to tack-free state at 180°F., in min.	(1)	(1)	240	(3)	120	65	150	150	30-35	20	40-60	64	51
Cure time for test sheets at 180°F., in hr.	16-22	22	16-22	22	16-22	22	16-22	16-22	22	22	22	22	22
Vulcanizate properties of test sheets:													
Hardness, in Shore "A" durometer degrees	(1)	(1)	39	27	36	31	40	41	34	33	34	36	35
Ultimate tensile strength, in percent	(1)	(1)	242	95	209	142	180	202	165	158	163	189	195
Ultimate elong., in percent	(1)	(1)	290	440	280	280	240	260	310	330	350	290	390

1No cure
2Recipe contains 0.17 pbw of Duponol L-144-WPG.
3Not tested.

One, that containing cupric chloride and water, was blended on a paint mill according to the recipe:

Curing Paste 1

Ingredient	Parts by Wt.
MnO_2	47.6
HB-40 terphenyl	37.5
Sterling MT carbon black	11.9
$CuCl_2 \cdot 2H_2O$	2.38
Water	2.38
(Copper content of paste)	0.89

and was used in Examples 10 to 13, 15 and 16. The other curing paste, that containing cupric acetate and water, was blended on a paint mill according to the recipe:

Curing Paste 2

Ingredient	Parts by Wt.
MnO_2	48.3
HB-40 terphenyl	35.4
Sterling	11.8
$Cu(C_2H_3O_2)_2 \cdot H_2O$	1.1
Water	4.4
(Copper content of paste 2)	0.35

and was used in Example 14. The copper salts and water were added to the other curing paste ingredients in the form of solutions prior to milling. Again, all ingredients in the table of examples are listed as separate entities, rather than as components of partial mixtures.

The inconsistent and unstable curing qualities and vulcanizate properties of prior art compositions are to be seen upon comparison of Examples 4 to 9. They appear to evidence an unusual sensitivity to changes in recipe amounts and components. Compositions of this process, Examples 10 to 16, on the other hand, show enhanced consistency, stability and lack of undue sensitivity in these regards despite changes in amounts of recipe ingredients.

BENZOTHIAZOLE-2-SULFENAMIDE CURE ACCELERATORS

A process developed by E.G. Millen and P.A. Koons (U.S. Patent 3,487,052; December 30, 1969; assigned to Thiokol Chemical Corporation) cures liquid polymers containing —SH groups with inorganic oxidizing agents in the presence of benzothiazole-sulfenamide type compounds which have an accelerating or synergistic effect on the curing reaction.

The benzothiazole-2-sulfenamides type compounds correspond to the general formula shown on the following page.

96

wherein R and R' may be the same or different and when taken singly are selected from the group consisting of hydrogen, alkyl, branched alkyl and cycloalkyl groups and when taken collectively with the nitrogen atom to which they are attached, form a heterocyclic group selected from the group consisting of: (a) azahydrocarbon, (b) azaoxahydrocarbon, (c) azathiahydrocarbon, and (d) azaoxathiahydrocarbon groups. Examples of the above hetero-cyclic groups are:

and

The benzothiazole cocuring agents of the process may have substituents such as halogen, alkyl, nitro groups, etc. on the benzene or heterocyclic ring. The organic sulfur-contain-ing cocuring agents may each be used singularly or in various combinations with one another.

Examples 1 to 4:   In Examples 2 to 4, a group of thiazole and thiazoline compounds were each used at two cure temperatures as a cocuring agent in combination with zinc peroxide to cure a liquid polysulfide polymer, designated as LP-31. The polymer is essentially linear with a small amount of branching or cross-linking. It has an average weight of 7,500 and a viscosity of 800 to 1,400 poises at 25°C.

Example 1 is a control using only zinc peroxide as the curing agent. In each case, the thiazole compound accelerated the cure of the polymer as compared with the control. The cure results are listed in the table below. The cure formation used in each case, by weight, was 25 parts of LP-31, 1.5 parts of zinc peroxide and 1 part of the cocuring agent.

Example	Co-curing agent	Cure results obtained at—	
		75° F.	180° F.
1	Control	Soft cure in 12 days	Soft cure in 7 days
2	2-mercaptothiazoline	Soft cure in 8 days	Cure in 2 hrs.
3	1,3-bis(2-benzothiazolyl-mercaptomethyl) urea.	No cure	Soft cure in 4 hrs.
4	2-mercaptobenzothiazole	Soft cure in 5 days	Cure in 1 hr.

Example 5 and 6:   In Example 6, Zetax was used as a cocuring agent with barium manganate to cure LP-31 liquid polysulfide polymer. Example 5 is the control using only barium manga-nate as the curing agent. The cure formulation used was, by weight, 25 parts of LP-31, 2 parts of barium manganate and 2 parts of the zinc salt of 2-mercaptobenzothiazole.

Example	5	6
Cure results obtained at 75°F.	No cure in 30 days.	Soft cure in 14 days.
Cure results obtained at 180°F.	No cure in 7 days.	Soft cure in 4 hours.

Examples 7 to 10: In Examples 8 and 10, two curing systems were used to cure a liquid —SH terminated polypropylene glycol polymer having a molecular weight of approximately 2,000 to 3,000. Examples 7 and 9 are controls in which only the oxidizing agent was used as a curing agent for the polymer used in Examples 8 and 10. In both cases, the cocuring agent exhibited an accelerating or synergistic effect as compared with the cure using the oxidizing agent alone. The formulations given in the table are presented in parts by weight.

Example	7	8	9	10
Formulation:				
Polymer	25	25	25	25.
Li₂O₂	2	2		
KIO₃			2	2.
Benzothiazolyl disulfide		2		
N-oxydiethylenebenzothiazole-2-sulfenamide				2.
Cure results obtained at 75° F	No cure in 5 days.	Soft cure in 1 hr.	No cure in 5 days.	Cure in 3 hrs.
Cure results obtained at 180° F	Soft cure in 1 day.	Cure in ½ hr.	Cure in 5 days.	Cure in ½ hr.

Examples 11 to 14: In Examples 12 and 14, two of the curing systems were used to cure a liquid —SH terminated butadiene/acrylonitrile copolymer. This copolymer contains about 24% acrylonitrile and has a viscosity of 35,000 cps and a specific gravity of 0.98 at 25°C.

Examples 11 and 13 are controls in which only the oxidizing agent was used as the curing agent for the same copolymer. In all cases the cocuring agent exhibited an accelerating or synergistic effect as compared with the cure using the oxidizing agent alone. The formulations given in the table are presented in parts by weight.

Example	11	12	13	14
Formulation:				
Copolymer	25	25	25	25.
(NH₄)₂Cr₂O₇	2	2		
CaO₂			2	2.
1,3-bis(2-benzothiazolylmercaptomethyl) urea		2		
N-cyclohexylbenzothiazole-2-sulfenamide				2.
Cure results obtained at 75° F	No cure in 7 days.	No cure in 7 days.	Soft cure in 23 days.	Soft cure in 5 days.
Cure results obtained at 180° F	Soft cure in 7 days.	Cure in 7 days.	Soft cure in 3 days.	Cure in 5 hrs.

Examples 15 to 25: A filled liquid polysulfide polymer system was tested with various combinations of cocuring agents as shown in the next two tables. The polymer was designated as LP-32. It has an average molecular weight of 4,000 and a viscosity of 350 to 450 poises at 25°C.

In examples 15 to 20, 100 parts by weight of LP-32 were mixed with 30 parts by weight of calcium carbonate, 10 parts by weight of titanium dioxide, 3 parts by weight of precipitated, hydrated silica, 10 parts by weight of ZnO_2 and 2 parts by weight of cocuring agents shown in the table. The control contained ZnO_2 as the only curing agent.

Ex.		Cure results obtained at 75° F.
15	Control	No cure in 7 days.
16	Benzothiazolyl disulfide	Cure in 4 days.
17	2-mercaptobenzothiazole	Cure in 4 days.
18	N-oxydiethylenebenzothiazole-2-sulfenamide.	Cure in 1 day.
19	N-cyclohexylbenzothiazole-2-sulfenamide	Cure in 1 day.
20	N,N-diisopropylbenzothiazole-2-sulfenamide.	Cure in 4 days.

In Examples 21 to 25, 100 parts by weight of LP-32 were mixed with 30 parts by weight of calcium carbonate, 10 parts by weight of titanium dioxide, 3 parts by weight of precipitated, hydrated silica, an inorganic oxidizing agent and 2 parts by weight of a thiazole compound as cocuring agent as shown in the table.

Ex.	Thiazole compound	Oxidizing agent (parts by weight)	Cure results obtained at 75° F.
21	None	MnO_2 (6)	Soft cure in 4 days.
22	N-oxydiethylenebenzo-thiazole-2-sulfenamide.	MnO_2 (6)	Soft cure in 1 day.
23	N-cyclohexyl benzo-thiazole-2-sulfenamide.	MnO_2 (6)	Cure in 2 days.
24	None	TeO_2 (8)	Cure in 1 day.
25	N-oxydiethylenebenzo-thiazole-2-sulfenamide.	TeO_2 (8)	Cure in 3 hrs.

In each of Examples 16 to 20, 22, 23, and 25, the combination of oxidizing agent with a thiazole cocuring agent produced a faster cure at 75°F. than did the oxidizing agent alone.

CUPRIC ABIETATE-CUMENE HYDROPEROXIDE CURE SYSTEM

In a process developed by J.R. Panek and O. Lamboy (U.S. Patent 3,505,258; April 7, 1970; assigned to Thiokol Chemical Corporation), cupric abietate is added to a cumene hydroperoxide cure system for —SH terminated liquid polysulfide polymer in trace amounts and in advantageous ratios relative to the amount of cumene hydroperoxide used per 100 parts of polymer thereby obtaining an extended, predictable and controllable range of working lives for precured mixtures of the polymer and the cure system ingredients, while also providing rapid cure times.

Examples 1 to 23: The process is illustrated with two commercially available —SH terminated polysulfide polymers of different molecular weights which require different amounts of cumene hydroperoxide curing agent to become cured, thus showing the effects of the cupric abietate regulating agent at two levels of addition. One of the polymers used is Thiokol LP-31, the other is Thiokol LP-32.

The two polymers were used in the form of masterbatches, as shown in the table, having incorporated therein other ingredients commonly used in the preparation of a typical coating composition having adhesive properties. In the masterbatches, the Methylon 75108, a phenolformaldehyde condensate, contributes adhesive properties to the LP-31 and LP- 32 polymer compositions. The aroclor 1262, a chlorinated biphenyl dissolved in an ester plasticize, contributes plasticizing properties, as does the sulfur. The other ingredients serve principally as fillers.

Masterbatch	A	B
Recipe ingredients in parts by weight (pbw.):		
LP-31 polysulfide polymer	100	......
LP-32 polysulfide polymer	......	100
Methylon 75108 (phenol-formaldehyde condensate)	5	5
Aroclor 1262, (chlorinated biphenyl, 60/40 in ester plasticizer)	35	35
Multifex MM (precipitated CaCO₃)	15	15
Icecap K (calcined clay)	20	20
York White (ground clay)	16	16
TiO₂	10	10
Hi Sil 233 (hydrated silica)	7	7
Sulfur	0.1	0.1
Total	208.1	208.1

A series of mixed curing pastes was then prepared using as the ingredients of each paste cumene hydroperoxide and cupric abietate and sufficient Aroclor 1254, a chlorinated biphenyl, to dissolve the cupric abietate. The curing pastes then were each mixed with portions of the two masterbatches in the quantities shown in Examples 1 to 23 of the next two tables, using 100 parts of polymer as a basis in each example. The first table shows results obtained with LP-31 polymer, and the second shows results obtained with LP-32 polymer.

Example	208.1 pbw. Masterbatch (100 parts) (polymer)	Curing paste, pbw. I. Cupric abietate[1]	II. Cumene hydroperoxide	Ratio I:II	Working life, hours	Curing time, hours at 75° F.	Durometer Hardness After 1 hour at 158° F.	After 5 hours at 158° F.
1	A	0.0012	4.8	25×10^{-5}:1	5	16	8	C
2	A	.0006	4.8	12.5×10^{-5}:1	6	16	8	C
3	A	.00012	4.8	2.5×10^{-5}:1	10	PC, 7 days	4	C
4	A	.00006	4.8	1.25×10^{-5}:1	NC	do	0	NC
5	A	.0010	4.0	25×10^{-5}:1	6	16	8	C
6	A	.0005	4.0	12.5×10^{-5}:1	15	PC, 7 days	0	PC
7	A	.0001	4.0	2.5×10^{-5}:1	15	do	0	PC
8	A	.00005	4.0	1.25×10^{-5}:1	NC	do	0	NC
9	A	.0008	3.2	25×10^{-5}:1	8	do	0	NC
10	A	.00008	3.2	2.5×10^{-5}:1	20	do	0	NC
11	A	.00004	3.2	1.25×10^{-5}:1	NC	do	0	NC

[1] 1% solution in Aroclor 1254 (chlorinated biphenyl).
C=Cured; NC=Not cured; PC=Partially cured.

Example	208.1 pbw. Masterbatch (100 parts) (polymer)	Curing paste, pbw. I. Cupric abietate[1]	II. Cumene hydroperoxide	Ratio I:II	Working life, hours	Curing time, hours at 75° F.	Durometer Hardness After 1 hour at 158° F.	After 5 hours at 158° F.
12	B	0.0024	9.8	24.5×10^{-5}:1	3	16	13	C
13	B	.0012	9.8	12.25×10^{-5}:1	4	16	11	C
14	B	.00024	9.8	2.45×10^{-5}:1	10	16	10	C
15	B	.00012	9.8	1.225×10^{-5}:1	9	16	11	C
16	B	.0020	8.0	25×10^{-5}:1	4	16	13	C
17	B	.0010	8.0	12.5×10^{-5}:1	5	16	10	C
18	B	.0002	8.0	2.5×10^{-5}:1	15	24+	6	C
19	B	.0001	8.0	1.25×10^{-5}:1	15	24+	6	C
20	B	.0016	6.4	25×10^{-5}:1	5	16	7	C
21	B	.0008	6.4	12.5×10^{-5}:1	12	16	4	C
22	B	.00016	6.4	2.5×10^{-5}:1	24+	24+	PC	PC
23	B	.00008	6.4	1.25×10^{-5}:1	24+	PC, 7 days	PC	PC

[1] 1% solution in Arocolor 1254 (chlorinated biphenyl).
C=Cured; NC=Not cured; PC=Partially cured.

Curing

After mixing the ingredients in the usual manner, each mixture consisting of the master-batch, curing agent and regulating agent ingredients was observed over a period of hours to determine the working life, or "pot-life" and the cure time. During this observation period, each mixture was observed as it thickened from a workable consistency, to a heavy syrup, or to a gelled, or set condition of unworkable consistency at room temperature, about 75°F. Portions of each batch were applied as coatings to aluminum, glass and wet concrete test panels for curing testing at 75°F. and at 158°F.

Adhesions of the cured compositions generally were of the same peel strengths as those usually obtained with the cumene hydroperoxide cured masterbatches in the absence of the cupric abietate, showing no adverse effects resulting from the use of the cupric abietate regulating agent. Since the process concerns principally the advantages in extending work-ing life of the precured polymer the test data shown in the above two tables have been con-fined to such aspect of the test results and to the cure rates.

Examples 1 to 11 show results obtained with amounts of cupric abietate ranging from 0.00004 part to 0.0008 part as 3.2 parts of cumene hydroperoxide to from 0.00006 part to 0.0012 part of cupric abietate at 4.8 parts of cumene hydroperoxide, all at 208.1 parts of masterbtach A, containing 100 parts of LP-31. Similarly, Examples 12 to 23 show results obtained with amounts of cupric abietate ranging from 0.00008 to 0.0016 part at 6.4 parts of cumene hydroperoxide to from 0.00012 part to 0.0024 part of cupric abietate at 9.8 parts of cumene hydroperoxide, all at 208.1 parts of masterbatch B, containing 100 parts of LP-32.

PROCESS CONTROL

CONTROL OF FREE THIOL UNITS

Three or More Free Thiol Groups

The oxidation of dithiols results in the formation of solid linear polysulfide polymers which have very little power of recovery, possess the disadvantage of cold flow and are soluble in many of the commonly used organic solvents, e.g., ethylene dichloride, chloroform, trichloroethane, etc. By bridging the linear chains with cross connecting links or groups of atoms, the power of recovery, resistance to cold flow and insolubility in usual organic solvents can be imparted to the polymers. A method for making cross-linked polysulfide polymers from polythiols is to incorporate with a dithiol starting material a higher polythiol containing from three to four or more thiol groups in the molecule in the desired proportion and subject the mixture to oxidation to obtain a solid polysulfide polymer. However, polythiols containing three or more thiol groups in the molecule are difficult to manufacture and are costly.

A process developed by E.D. Holly (U.S. Patent 2,831,896; April 22, 1958; assigned to Dow Chemical Company) gives mixtures of polythiols in liquid form and so reactive that upon suitable treatment they may be converted into insoluble cross-linked polysulfide polymers. It also provides a method of converting a portion of a dithiol starting material into a higher polythiol containing at least three thiol groups in the molecule to obtain a reactive liquid composition comprising essentially a mixture of the dithiol starting material and a higher polythiol. Furthermore, it offers a method of making such reactive polythiol liquid compositions containing a predetermined and controllable number of thiol units.

A portion of a dithiol starting material can be converted to the corresponding monoalkali metal derivative by reacting the dithiol with less than a chemically equivalent amount of an alkali-acting material such as an alkali metal, e.g., sodium, potassium or lithium, an alkali metal hydroxide or an alkali metal alcoholate. The reaction which occurs readily at room temperature or thereabout is usually carried out in a liquid reaction medium comprising a lower aliphatic alcohol or an aqueous solution containing the alcohol in a concentration of 50% by weight or greater.

The reaction medium is employed in an amount sufficient to substantially maintain the reactants in solution or form a mixture or slurry which can conveniently be stirred. The reaction medium may be used in as large a proportion as desired. The alkali-acting material, i.e., the alkali metal, the alkali metal hydroxide or the alkali metal alcoholate can be employed in amounts from 0.01 to 0.5, gram atomic proportion of alkali metal per gram molecular equivalent proportion of the dithiol starting material. The monoalkali metal derivative of the dithiol is usually prepared by dissolving the dithiol starting material in an equal volume of alcohol, e.g., ethyl alcohol (95%), and adding the alkali-acting material such as sodium metal, sodium hydroxide or sodium alcoholate thereto with stirring in the desired proportion.

The monoalkali metal derivative of the dithiol is reacted with a polyhalo organic compound such as the cyclic trimer of chloroacetaldehyde or an acyclic or an alicyclic polyhalohydrocarbon in the solution in which it is prepared, i.e., in the presence of the remaining portion of the dithiol starting material and the liquid reaction medium in which it was prepared, by adding to the mixture an amount of the polyhalohydrocarbon corresponding approximately to one gram atomic proportion of halogen per gram atomic proportion of alkali metal in the mixture. The resulting mixture is maintained at reaction temperatures preferably from 60° to 110°C. and under a pressure at least sufficient to maintain the mixture substantially in liquid phase under the conditions employed. The reaction can be carried out at atmospheric or superatmospheric pressures.

The reaction is usually complete in a period of from 10 minutes to 24 hours or longer, depending for the most part upon the temperature employed. The mixture should not be heated at elevated temperatures for prolonged periods of time or at reaction temperatures substantially greater than about 120°C. to avoid deterioration or condensation of the polythiol product which results in partial polymerization of the product to form polysulfide polymers. The reaction is carried out in the absence or substantial absence of air or oxygen, although the reaction can be carried out in the presence of a limited amount of air such as under reflux.

Upon completing the reaction, the mixture is heated under subatmospheric pressures to distill and separate the lower boiling liquid reaction medium, e.g., alcohol, from the product. The residue or liquid product is separated from the by-product solid material by decanting or filtering. The crude reaction mixture, if alkaline, is usually treated with an acid such as hydrochloric acid to bring the liquid to a pH value between 5 and 6 prior to recovering the polythiol product, but such procedure is not required.

The product, i.e., the liquid mixture of polythiols, is useful for a variety of applications. It can be cured to transform the liquid to the solid condition to form insoluble cross-linked polysulfide polymers. Oxidation in one step is used for that purpose. The product may be mixed with liquid polysulfide polymers having a molecular weight of from about 500 to 15,000 and the mixture cured or oxidized to the solid condition to form rubber-like polysulfide polymers which are useful for a variety of applications. The product may be incorporated with solid polysulfide polymers to facilitate processing or to secure the advantages of liquid polysulfide polymers over solid polymers.

Tri- and Tetramercapto Compounds

M.B. Berenbaum, J.R. Panek and L. Citarel (U.S. Patent 3,402,134; September 17, 1968; assigned to Thiokol Chemical Corporation) have developed another process for controlling the number of free thiol groups where liquid polysulfides have added to them prior to curing about 0.05 to 20 parts by weight of one or more polymercaptan cross-link agents of the type R'(SH)$_m$ where m is a whole number of 3 or 4, and R' is an organic radical that is inert with respect to reaction with mercaptan groups, such alkyl or aryl hydrocarbon, ether, polyether, ester, polyester, amide or similarly inert radicals, per one hundred parts by weight of the liquid polysulfide polymer in such compositions. The resulting compositions which also contain suitable quantities of curing agents, may then be cured at temperatures of about 65° to 300°F. to produce elastomeric vulcanizates having selected cross-link dependent properties.

The polymercaptan cross-link agents found useful in the practice of the process may effectively be used with the liquid polysulfide polymers which previously have been prepared with no cross-link or with up to 4% cross-link. The amount of any specific polymercaptan cross-link agent to be used in the formulation of a curable polysulfide polymer based composition for any specific end use is dependent upon the type and extent of control needed over the cross-link dependent properties of the vulcanizate. In general, trimercaptan cross-link agents will provide a smaller degree of cross-link control per mol used than the corresponding tetramercaptan cross-link agents.

The process may be considered as essentially a two-step process which includes as the first step the mixture of at least the necessary component ingredients required to form the curable compositions. In the second step, the blended curable compositions are cured at temperatures of about 65° to 300°F. for about 1 minute to 72 hours to produce the desired vulcanizate.

<u>Examples 1 to 21:</u> In the following examples, an intimate blend of the listed ingredients of the curable compositions was prepared by milling together the polysulfide liquid polymer with the polymercaptan cross-link agent and at least a portion of the other adjuvants as one partial mixture, and by milling together the curing agent in a liquid plasticizer vehicle with perhaps some of the other adjuvants as the second partial mixture, then combining the two partial mixtures by hand mixing at room temperature to form the specific curable composition.

The cure time was observed to be that interval of time required, at a specified temperature, to obtain a tack-free elastomeric solid. In general, the compositions were cured to form elastomeric vulcanizates in the form of test sheets. These sheets were then examined for physical property data, such as tensile properties (ASTM D412-51T), hardness (ASTM D676-59T) and compression set (ASTM D395-55).

In Examples 1 to 4, polysulfide liquid polymer A has a linear backbone that is essentially HS(C$_2$H$_4$OCH$_2$OC$_2$H$_4$SS)$_n$C$_2$H$_4$OCH$_2$OC$_2$H$_4$SH, where n is such that the MW is about 4,000. Polymer A was made with no cross-linking agent and would cure in prior art compositions with a nominal zero percent cross-link. In Examples 5 to 13, a blend of two polysulfide liquid polymers, LP-12 polymer and LP-32 polymer, was used. LP-12 polymer is similar to polymer A in all respects except that it has a nominal cross-link of 0.1% provided by trichloropropane. LP-32 polymer is also similar to polymer A in all respects except that its nominal cross-link is 0.5%. The nominal cross-link of the blend is about 0.15%.

It is to be seen in Examples 1 to 13 that the cross-link dependent property "ultimate percent elongation" can be controllably and substantially changed by varying the amount of the polymercaptan cross-link agent used in the various compositions.

104

Example	1	2	3	4	5	6	7
Recipe of Curable Composition, in pbw.:							
Polysulfide liquid polymer A	100	100	100	100			
LP-12 Polysulfide liquid polymer					75	75	75
LP-32 Polysulfide liquid polymer					25	25	25
Trimethylolpropane tris(3-mercaptopropionate)							
Pentaerythritol tetrakis (3-mercaptopropionate)	0	0.1	0.5	1.0	0	0.1	0.5
Stearic acid	1	1	1	1	0.5	0.5	0.5
Semi-reinforcing furnace grade carbon black	30	30	30	30	30	30	30
PbO$_2$, 50% in Aroclor 1254 liquid vehicle	15	15	15	15	15	15	15
Curing Conditions:							
Cure interval, in minutes	[1]	[1]	[1]	[1]	10	10	10
Cure temperature, in °F	80	80	80	80	225	225	225
Vulcanizate Properties:							
Tensile strength, in p.s.i.	398	418	326	310	125	120	170
Ultimate elongation, in percent	970	860	770	560	>1,100	760	680
Hardness, in Shore A durometer degrees	49	48	51	59	32	33	38

Example	8	9	10	11	12	13
Recipe of Curable Composition, in pbw.:						
Polysulfide liquid polymer A						
LP-12 Polysulfide liquid polymer	75	75	75	75	75	75
LP-32 Polysulfide liquid polymer	25	25	25	25	25	25
Trimethylolpropane tris(3-mercaptopropionate)			0.5	1.0	2.0	5.0
Pentaerythritol tetrakis (3-mercaptopropionate)	1.0	5.0				
Stearic acid	0.5	0.5	0.5	0.5	0.5	0.5
Semi-reinforcing furnace grade carbon black	30	30	30	30	30	30
PbO$_2$, 50% in Aroclor 1254 liquid vehicle	15	15	15	15	15	15
Curing Conditions:						
Cure interval, in minutes	10	10	10	10	10	10
Cure temperature, in °F	225	225	225	225	225	225
Vulcanizate Properties:						
Tensile strength, in p.s.i.	200	185	205	225	230	195
Ultimate elongation, in percent	430	80	920	680	310	125
Hardness, in Shore A durometer degree	42	55	38	42	49	52

Example	14	15	16	17	18	19	20	21 [3]
Recipe of Curable Composition, in pbw.:								
LP-3 Polysulfide liquid polymers	100	100	100	100	100	100	100	100
Trimethylolpropane tris(3-mercaptopropionate)		1.5						
Trimethylolethane tris(3-mercaptopropionate)				1.5	4.5			
Pentaerythritol tetrakis (3-mercaptopropionate)	1.5		4.5					0.3
MnO$_2$, 50% in Aroclor 1254 liquid vehicle	7.5	7.5	7.5	7.5	7.5	7.5		
Semi-reinforcing furnace grade carbon black	45	45	45	45	45	45	30	30
Dinitrobenzene	7.5	7.5	7.5	7.5	7.5	7.5		
O-nitroanisole							12	12
Curing Conditions:								
Cure interval, in minutes	[1]	[1]	[1]	[1]	[1]	[1]	[1]	[1]
Cure temperature, in °F	80	80	80	80	80	80	80	80
Vulcanizate Property:								
Compression set, at 80° F., in percent	9	4	10	13	8	25	[2]	1.8
Compression set, at 158° F., in percent	95	43	100	100	87	100	[2]	40

[1] Not measured. [2] No cure. [3] 0.17 pbw. of Duponol L–144–WDG, a sodium salt of a modified unsaturated long chain alcohol sulfate was added to the recipe prior to cure.

In Examples 14 to 21, LP-3 polysulfide liquid polymer was used, which polymer is similar to liquid polymer A except that it has an average molecular weight of about 1,000 and a cross-link of 2% imparted by trichloropropane. These examples show that the use of poly-mercaptan cross-link agents in the curable compositions provide control over compression set properties of vulcanizates made therewith, and especially when the curing is conducted at ambient temperatures. Such ambient temperature cures usually require several days time. The manganese dioxide used in these examples was Manganese Hydrate No. 37, a substance being 70% MnO$_2$, 22% water of hydration, 43.8% manganese and 12.5% in available oxygen. The solvent vehicle, Aroclor 1254, is a polychlorinated biphenyl containing 54% chlorine.

FERRIC SALT AND AMMONIA OR AMINE AS POLYMERIZATION CATALYST

In a process developed by W.R. Nummy (U.S. Patent 2,866,776; December 30, 1958; assigned to Dow Chemical Company) di(2-mercaptoethyl) ethers of diols can readily be polymerized by oxidizing the compounds with air or oxygen in the presence of small but effective amounts of both an inorganic iron salt, e.g., ferric chloride or ferric sulfate, and ammonia or a primary amine, preferably in mixture with trace amounts of water or water vapors. It has been found that carrying out of the condensation or oxidation in the presence of both an inorganic iron salt and ammonia or a primary amine results in a faster rate of reaction than is obtained in carrying out of the reaction in the presence of either the iron salt or ammonia or a primary amine alone. The employment of both an iron salt and ammonia or a primary amine appears to have a synergistic action for catalyzing the condensation or oxidation reaction and results in substantial improvement for effecting the condensation over that obtained by carrying out of the reaction with either of the agents alone under otherwise similar reaction conditions.

The iron salts to be employed as catalysts or promoters for the condensation reaction are inorganic salts which are soluble in the di(mercaptoethyl) ethers of diols starting materials. Examples of such iron salts are ferric chloride, ferric bromide, ferric nitrate, ferric sulfate, ferrous sulfate, ferrous chloride or ferrous bromide. The iron salts can be employed in amounts preferably from 0.001 to 0.01% by weight of the starting material. With smaller amounts of the iron salt, the condensation reaction proceeds at a slow rate. Larger proportions of the iron salt result in the formation of a polymer product of darker color than is usually desired.

The iron salts are preferably used in anhydrous or substantially anhydrous condition, although the iron salts containing water of crystallization such as $FeBr_3 \cdot 6H_2O$, $FeCl_3 \cdot 6H_2O$, $Fe(NO_3)_3 \cdot 9H_2O$, $FeSO_4 \cdot 7H_2O$ or $FeCl_2 \cdot 4H_2O$ can be used. The inorganic salts are employed, together with a nitrogen-containing base such as ammonia or a primary amine.

The ammonia or primary amine can be used in amounts preferably from 0.05 to 1% by weight of the di(2-mercaptoethyl) ether of a diol starting material employed. The ammonia or primary amine can be mixed with the di(2-mercaptoethyl) ether of diol starting material or with a stream of air or oxygen gas fed into mixture with the liquid starting material containing the iron salt in the desired proportion. For convenience in handling, the volatile nitrogen-containing bases, such as ammonia or methyl amine, can readily be added to the mixture by bubbling a stream of air or oxygen through an aqueous solution of the same and feeding the moist vapors into mixture with the di(2-mercaptoethyl) ether containing the iron salt.

When the amine is a liquid or solid which can conveniently be handled, it is preferably added directly to the di(2-mercaptoethyl) ether of diol starting material, as is the inorganic iron salt. Mixtures of any two or more of the inorganic iron salts and mixtures of ammonia and one or more primary amines or mixtures of any two or more primary amines can be used as the catalyst materials for promoting the condensation of di(2-mercaptoethyl) ethers of diols having the aforementioned general formula in the presence of oxygen or air to form polysulfide resins as herein described.

The catalyst materials, i.e., the iron salt and the ammonia or primary amine, are most effective in promoting the condensation or oxidation reaction of the di(2-mercaptoethyl) ethers of diols to form polysulfide liquid polymers when employed in the presence of small amounts of water, e.g., from 0.1 to 2% by weight of the water based on the di(mercaptoethyl) ether of diol starting material. Usually an amount of water such as the water vapor carried in a stream of air or oxygen, obtained by bubbling the gas through a tank of water, is satisfactory. The reaction can be initiated without adding water to the mixture or to the stream of oxygen gas, although in such instance the rate of reaction is somewhat slower than when water is used. The oxidation reaction has not been carried out under truly anhydrous condition. However, since water is a by-product of the reaction it appears that a trace amount of water vapor facilitates the oxidation reaction. The proportion of water should not exceed the amount which can be dissolved in the mixture and lesser amounts are usually required. The water can be employed in amounts of from 0.05 to 5, preferably from 0.05 to 1% by weight of the starting material.

The condensation or oxidizing reaction can be carried out at temperatures preferably from 70° to 120°C., and at atmospheric or superatmospheric pressures. The reaction is somewhat sluggish at room temperature or thereabout, and at higher temperatures, e.g., at temperatures of from 150° to 200°C. within the range stated, results in a product of somewhat darker color than is obtained at reaction temperatures within the preferred temperature range, under otherwise similar reaction conditions.

In a preferred practice, the di(2-mercaptoethyl) ether of a diol starting material, together with the inorganic iron salt such as anhydrous ferric chloride or ferric bromide, are mixed together in the desired proportions. The mixture is agitated and maintained at a reaction temperature within the range of from 70° to 120°C., and air or oxygen gas, suitably saturated with water vapor, and in mixture with ammonia or a volatile primary amine, e.g., methylamine, in the desired proportion is fed into contact or mixture with the liquid di(2-mercaptoethyl) ether of a diol starting material containing the iron salt suitably at atmospheric or substantially atmospheric pressure. The reaction is continued until the desired degree of condensation or polymerization of the compound is attained, suitably until the mixture has an absolute viscosity corresponding to from 500 to 100,000 centipoises at 25°C., and prior to gelling of the mixture. The reaction is discontinued by stopping feed or contact of the air or oxygen with the mixture.

The resulting product is a liquid polysulfide resin containing reactive mercapto groups and can readily be oxidized, e.g., by treatment with a further amount of oxygen or air, or cured by substances which condense with the hydrogen of the mercapto groups and act as curing agents, to form normally solid rubbery materials. For these reasons the liquid polysulfide resins are maintained in the absence or substantial absence of air or oxygen prior to their final use in the preparation of an end product.

LOW-TEMPERATURE POLYMERIZATION

A process developed by B.W. Hubbard (U.S. Patent 3,164,643; January 5, 1965; assigned to Ideal Roller and Manufacturing Company) gives polysulfides having especial utility in the lithographic industry and one made by rapid, low temperature reactions. While the application of the compositions to the lithographic field, e.g., as roller surfaces and in blankets, is a very important one, these compositions are generally useful wherever tough, chemically resistant, resilient compositions are desired.

Briefly, the compositions are prepared by subjecting a liquid polysulfide polymer (a useful commercial product is that marketed under the trade name "Thiokol Liquid Polymer LP-3") to reaction conditions in the presence of a minor amount of an oxide of an element of

Group VIa of the Periodic Table, for example, the oxides of tellurium or selenium, and a primary or secondary amine. There is also added to the reactants for the purpose of reinforcing the ultimate product an amount of a filler such as titanium dioxide or equivalent powder, e.g., zinc oxide, carbon black, etc., and a minor amount of a material capable of assuring uniform dispersion of the filler throughout the reaction mix and final product. For this purpose an epoxide polymer or a chlorinated hydrocarbon, e.g., Aroclor 1254, has been found especially useful. A relatively minor amount of m-dinitrobenzene or equivalent solvent should be added to the reaction mix for the purpose of lowering the surface tension of the mix to permit entrapped gases or air to readily be released and thereby decrease porosity of the finished product.

A mixture of polysulfide, titanium dioxide, dinitrobenzene and epoxy resin, which may be referred to as Part A of the reaction mixture, is first prepared. In Part A the polysulfide to epoxide ratio comprises about 33 parts by weight to 1, the titanium dioxide comprises a little more than one-tenth the weight of polysulfide and the m-dinitrobenzene is about one-third the weight of titanium dioxide. The tellurium dioxide, in an amount about equal to that of the epoxy resin in Part A is mixed with a minor amount of a suitable dispersing liquid for the tellurium dioxide. Dibutylphthalate has been found to be particularly useful as a dispersing liquid but any equivalent material may be employed if desired.

Part B, i.e., the tellurium dioxide and dibutylphthalate is equal to about one-tenth the weight of Part A with which it is mixed. These two parts may be combined without initiation of the exothermic reaction which characterizes the ultimate formation of the elastomers. Indeed, it is preferred that Part A and Part B be thoroughly mixed some time prior to that when it is desired to effect the cure. A mixture of A and B without amine has a stable life of as much as 5 to 8 hours. Once the amine is added to the mixture of Parts A and B the exothermic reaction begins almost immediately and goes rapidly to completion at room temperature without application of additional heat. The reaction is usually completed in 5 to 10 minutes or less.

Example: Based upon the final reaction mixture, including the polysulfide, epoxy, amine, oxide, filler, solvents, etc., Part A of the final reaction mixture was first prepared comprising 75.20 parts by weight of Liquid Polymer LP-3, 11.30 parts of titanium dioxide filler, 3.75 parts of m-dinitrobenzene and 2.25 parts of Epon 828. These components were mixed thoroughly and Part B was prepared comprising 2.47 parts of TeO_2 and 1.23 parts of dibutyl phthalate which were also mixed together to form a paste. Parts A and B were then mixed together preparatory to carrying out the cure.

The mold, a length of pipe about five feet in length and about five inches in diameter was placed in a special horizontal lathe and rotation was begun. A wax liner was first laid down on the inside of the mold by introducing molten paraffin wax into one end while applying cooling water to the exterior of the mold. 3/8 part of tertiary butyl amine were then mixed with the mixture of Parts A and B and the combined reactants were fed into one end of the rotating mold. The cure was completed in 5 to 10 minutes with a shell of rather dense elastomer formed on the wax within the mold. Ice water cooling on the exterior of the mold was used to cool the product which had been heated by the exothermic curing reaction. When cure was complete, the mold was removed from the lathe, stood erect and a steel shaft was centered therein. At this stage there was mixed with a second batch of the mixture of

Parts A and B an amount of about 2% of tertiary butyl amine based upon the total composition. This reaction mixture was then poured in the open top of the mold (the bottom being sealed) between the outer elastomer shell and the shaft. Upon completion of the cure, heat was applied to the exterior of the mold, the wax was melted and the roll removed.

CONTROLLED MOLECULAR WEIGHT

A process developed by P.F. Warner (U.S. Patent 3,219,638; November 23, 1965; assigned to Phillips Petroleum Company) results in polysulfide polymers of controlled properties, particularly molecular weight, by the oxidation of dimercaptans with sulfur in the presence of a monomercaptan compound. Monomercaptans having up to and including 16 carbon atoms are effective for controlling the properties, especially molecular weight, of polysulfide polymers formed by contacting an organic dimercaptan compound with sulfur as an oxidizing agent in the presence of a basic catalyst. The polymers obtained are saturated and the linkage is through sulfur rather than olefin as is the case with many polymers. The molecular weight of polysulfide polymers obtained by the above reaction can be controlled by regulating the ratio of dimercaptan to monomercaptan during contacting.

Numerous variations in operative procedure can be employed. Ordinarily the monomercaptan and dimercaptan are contacted with each other prior to contacting with free sulfur or a sulfur-yielding compound and a basic catalyst. However, in some instances it may be desirable to charge all of the reactants simultaneously to a reaction zone. In any event, the reaction between the dimercaptan and sulfur is carried out in the presence of a monomercaptan. The reaction can be effected in a batch, intermittent or continuous manner.

The properties of the polymer products obtained will vary appreciably. The amount of sulfur present in the polymeric product will generally range from about 30 to about 60 weight percent. The polymers can be compounded by any of the known methods for compounding polysulfide polymers. Vulcanization accelerators, reinforcing agents and fillers such as have been normally employed in polysulfide rubbers can likewise be used in the compounds. The polymers have utility in applications where both natural and synthetic rubbers are used. For example, they can be used in the manufacture of automobile tires, gaskets and other rubber articles.

Example 1: A series of runs was carried out in which polysulfide polymers of 2,9-p-menthane dithiol were prepared by reacting 2,9-p-menthane dithiol with 2 mols of sulfur per mol of dithiol at a temperature of about 80°F. in the presence of diphenyl guanidine as a basic catalyst. In one run, the contacting was effected without a monomercaptan modifier and in three other runs n-butyl mercaptan and tert-dodecyl mercaptan were employed as polymerization modifier or polymer chain terminators. The contacting was effected at approximately one atmosphere pressure and in the presence of benzene as an inert diluent. In the runs employing monomercaptan modifier, the dimercaptan and monomercaptan were contacted with each other in the benzene diluent prior to contacting with sulfur and catalyst.

In these runs, benzene diluent, 2,9-p-menthane dithiol and either n-butyl mercaptan or tert-dodecyl mercaptan were charged together to a 2-liter reactor containing diphenyl guanidine catalyst. The reaction mixture thus formed was heated to a temperature of approximately

130°F. and then sulfur was added to the reaction mixture slowly over a period of about 30 minutes so as to keep the reaction mixture from boiling over due to rapid evolution of H_2S. After rapid evolution of H_2S had stopped, the benzene solution of polymer was heated to a temperature of approximately 200°F. at about 20 mm. Hg absolute pressure to remove the last traces of benzene solvent. The ratio of reactants employed in each run and properties of the resulting polymers are shown below.

	Run 1	Run 2	Run 3	Run 4
Modifier	None	(1)	(2)	(3)
Mol Ratio Dithiol/Monothiol	—	2.5	5	2.5
Mol Ratio Sulfur/Dithiol	2	2	2	2
Viscosity, poises at 200°F.	58	10	40	26
Mercaptan Sulfur, wt. percent	1.9	0.31	0.31	1.50
Molecular weight	1,214	1,052	1,497	854

(1) $n-C_4H_9SH$

(2) $n-C_4H_9SH$

(3) $tert-C_{12}H_{25}SH$

The table above shows the properties of sulfide polymers using mol ratios of dimercaptan to monomercaptan of 2.5 and 5 for n-butyl mercaptan and a mol ratio of dimercaptan to tert-dodecyl mercaptan of 2.5:1. As the amount of monomercaptan employed is increased there is a decrease in the viscosity of the resulting polymer.

Example 2: Control of the molecular weight of a polysulfide polymer is demonstrated by this example. The smallest repeating unit of a 2,9-p-menthane dithioltrisulfide polymer is as follows:

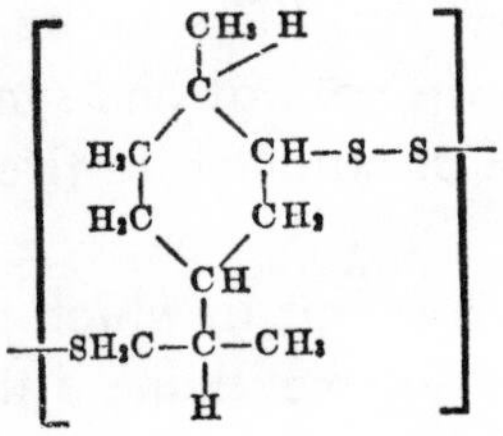

This unit has a molecular weight of 234. If it is desired to make a polymer having 10 repeating units, the following recipe is employed:

	Grams	Mols
2,9-p-menthane dithiol	2,340	10
n-Butyl mercaptan	180	2
Sulfur	640	20

The reactants are contacted in an inert hydrocarbon diluent at a temperature of about 80°F. with a basic catalyst (diphenyl guanidine) until the reaction is completed. The polymer is recovered from the reaction mixture and has the following structure:

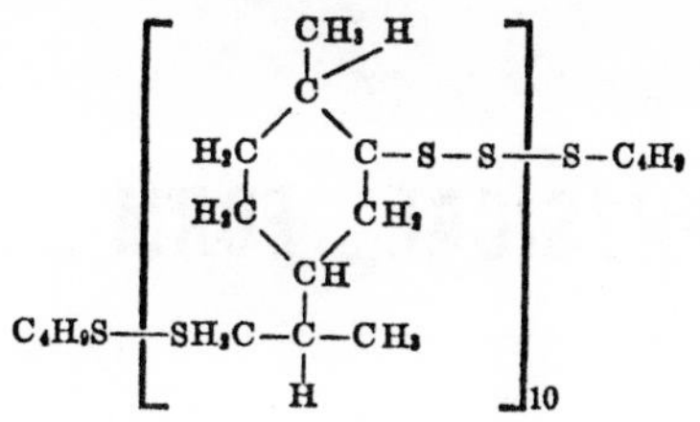

The above polysulfide polymer has a molecular weight of 2518. If a lower molecular weight polysulfide polymer is desired, the ratio of monomercaptan to dimercaptan is increased and when higher molecular weight sulfide polymers are desired the ratio is decreased. The polymers obtained above are clear resin-like solids at room temperature; however, they will cold flow over prolonged periods of storage.

PHYSICAL FORM

STABLE DISPERSIONS

Halogenated Paraffins/Polysulfides and Salt of H$_2$SO$_4$ Ester

Methods have been disclosed for the production of high molecular weight plastic reaction products by reacting a halogenated hydrocarbon, such as a dihalogen substituted paraffin, with an alkaline polysulfide, in an aqueous medium. By carrying out the reaction in the presence of a dispersing agent aqueous dispersions, or latices, resembling rubber latex are obtained. As dispersion agents such inorganic materials as magnesium hydroxide, obtained, for example, by the reaction of magnesium chloride with sodium hydroxide during the course of the reaction, have generally been employed. The dispersions thereby obtained are, however, generally handicapped by the presence of the dispersed reaction product in the form of relatively large grains or particle size. They are ordinarily stable only as long as the inorganic dispersion agent is present in relatively large amounts. Removal of the inorganic dispersion agent results in an immediate irreversible coagulation of the dispersed particles. Films formed from such dispersion are often highly unsatisfactory since upon evaporation of the aqueous medium therefrom a layer of crumbly character is generally formed lacking in a proper degree of adhesive power.

Utilization of dispersion agents of a more or less organic nature such as, for example, glue, soaps and the like, generally result in an unavoidable lack of uniformity of product and the presence therein of lumpy masses. Generally the dispersions so prepared can be freed of salt (NaCl) formed during the reaction, only with difficulty if at all. Due to the physical nature of a substantial part of the particles comprised therein, and the inability to effectively separate salt therefrom coherent films possessing suitable characteristics for practical use can generally not be produced therefrom by such means as subjection to evaporation.

Materials produced by a process developed by J. Overhoff and H.W. Huyser (U.S. Patent 2,469,226; January 31, 1950; assigned to Shell Development Company) possess improved dispersions of stable character by reacting a dihalogen substituted paraffin with an alkaline polysulfide in an aqueous medium in the presence of a water-soluble salt of a sulfuric acid ester of a monohydric C$_{10}$ to C$_{18}$ aliphatic alcohol.

Two Layer Reaction Product

Dispersions will often continue to undergo a settling out of the dispersed phase during storage resulting in their separation into two layers, a clear aqueous supernatant layer and a bottom layer, comprising the disperse phase. Forming the two layers into a single uniform dispersion, once they have been formed, by such practical means as stirring or the like, is generally exceedingly difficult if not impossible. The supernatant layer consists of an aqueous solution of the salt reaction product formed as a by-product during the reaction. When dichloroethane is reacted with an alkali metal polysulfide the salt by-product formed will be the chloride of the alkali metal. The supernatant aqueous layer formed may be separated from the lower layer by decantation. Since, however, the salt content of the supernatant layer is only a part of the total salt by-product a substantial amount of salt will remain in the bottom layer comprising the dispersed phase.

The instability of these dispersions is attributable to at least a substantial degree to the presence therein of the salt by-product. Consequently the bottom layer will continue to separate a clear supernatant aqueous salt solution from a bottom layer comprising the high molecular weight reaction product. Continued separation of a supernatant aqueous salt solution generally results in a conglomeration of particles of high molecular weight reaction product in the remaining bottom layer, which conglomerated particles are then finally converted into an irreversible coagulum.

Sedimentation of the dispersed particles with resultant formation of the supernatant aqueous salt solution is attributable to at least a substantial degree to the fact that the dispersed particles of high molecular weight reaction products are in a flocculant state in the aqueous medium of the dispersion in the presence of the salt concentration prevailing therein. This appears to be evidenced by the fact that upon dilution of the reaction mixture with copious amounts of water no stratification appears to take place, whereas restoration of the salt concentration to that prevailing in the reaction mixture generally obtained again enables stratification into a clear aqueous salt layer and a lower layer comprising the high molecular weight reaction products.

Another Shell process, this one developed by H. Eilers, J. Overhoff and J.C. Vlugter (U.S. Patent 2,496,227; January 31, 1950; assigned to Shell Development Company) gives improved highly concentrated aqueous dispersions of stable character by reacting a dihalogen substituted paraffin with an alkaline polysulfide in an aqueous medium in the presence of a dispersing agent, thereby forming high molecular weight plastic reaction products dispersed in an aqueous inorganic salt solution, adding an electrolyte providing a polyvalent cation to the dispersion thus obtained, thereby effecting a relatively rapid flocculation of the dispersed solids and a separation of the mixture into a supernatant layer comprising an aqueous salt solution and a lower layer comprising flocculated high molecular weight reaction products and the added polyvalent cation, separating the aqueous supernatant layer from the lower layer and removing the cation from the remaining lower layer thereby converting the lower layer into an irreversible highly concentrated aqueous dispersion.

Example 1: Dichloroethane was reacted with an aqueous 0.5 mol solution of sodium polysulfide at a temperature of 70°C. while stirring vigorously. A slight molal excess of dichloroethane over sodium polysulfide was employed. The reaction was executed in the presence of 0.3% by weight of the aqueous reaction mixture of a dispersing agent consisting essentially of a mixture of sodium salts of sulfuric acid esters of olefins having from 10 to 18 carbon atoms to the molecule (obtained by the vapor phase cracking of paraffin wax).

Due to the presence in the resulting reaction products of by-product sodium chloride in an amount sufficient to effect flocculation of the dispersed high molecular weight reaction product, a sedimentation of the high molecular weight reaction products was obtained with the separation of a clear supernatant aqueous salt layer. The supernatant aqueous salt layer was separated from the bottom layer by decantation.

To 100 kg. of the bottom layer thus formed, containing approximately 45 kg. of high molecular weight reaction products and approximately 2.5 kg. of sodium chloride, there was added 420 liters of water containing 420 grams of calcium chloride. Substantially complete flocculation of the high molecular weight reaction product occurred.

Upon standing, stratification of the mixture occurred with the formation of a supernatant layer of aqueous salt solution. The supernatant layer, consisting of about 410 liters, was separated from the bottom layer by siphoning. The the remaining bottom layer, consisting of about 140 kg. and containing now no more than approximately 0.5 kg. of sodium chloride, there was added 800 cc of 5% aqueous ammonium oxalate solution. A dispersion of high stability having a solid content in excess of 30% was thereby obtained containing calcium oxalate in finely divided dispersed state. The calcium oxalate in no wise adversely affected the stability or other desirable properties of the highly concentrated dispersion thus obtained.

Example 2: To a calcium chloride treated bottom layer obtained above, there was added another 450 kg. of water containing 400 kg. calcium chloride; a renewed flocculation of the high molecular weight dispersed particles and formation of a supernatant aqueous layer were thereby obtained. The supernatant layer thus formed was again removed by decantation and a bottom layer consisting of about 140 kg. was obtained comprising the high molecular weight reaction product in a flocculated state and containing now only about 0.125 kg. of chloride salts calculated as sodium chloride.

To the resulting bottom layer, twice treated with calcium chloride, thus obtained there was added 250 g. of sodium hexametaphosphate dissolved in 1 liter of water. A stable highly concentrated dispersion was thereby obtained having a solid content in excess of 30% by weight. The resulting dispersion thus obtained contained less chlorides than the stable dispersion above and was furthermore free of any crystalline phase such as calcium oxalate.

POWDER FORM

Polyethylene Polysulfide Containing 82% Sulfur

If a polymer could be recovered as a precipitate from the polymerization mass, it could be readily washed free of its acid content to obtain a dry powder, easily intermixed with vulcanizing and other agents and subsequently formed into a sheet or other product form. According to a process developed by H. Brückner and K. Weissörtel (U.S. Patent 2,906,739; September 29, 1959; assigned to Farbwerke Hoechst Aktiengesellschaft, Germany), polyethylene polysulfide containing approximately 82% sulfur, i.e., in theory $(C_2H_4S_4)n$, can be produced in powder form.

Example: Crystalline sodium sulfide in an amount of 400 kg. and 150 kg. of sulfur are dissolved in water while heating in accordance with known ways of producing sodium polysulfide to obtain 700 liters of solution. Upon the addition of more water, the solution is adjusted to a density of 1.2458 (a 2-molar solution). To this solution are added 17 kg. of 50% caustic soda and 22 kg. of crystalline magnesium chloride in 22 kg. of water. This mixture which now contains finely divided magnesium hydroxide is heated to about 65°C. and has 141 kg. ethylene dichloride (chromium number 100) added while stirring at a rate that the reaction temperature does not exceed 70° to 75°C. Upon the completion of the ethylene dichloride addition, the temperature is increased to 90°C. and stirring continued for another hour. Upon settling, the electrolyte containing supernatant liquid is drawn off and the residual solid contents diluted with water. Upon acidification with hydrochloric acid to a pH of 4.0, the resultant precipitated solid is washed on a suction filter and then is dried at a temperature of approximately 35°C. As a result, there is obtained a finely divided powdered polyethylene polysulfide product weighing approximately 210 kg. with a sulfur content of about 82%.

Redispersible Powder

The problem of producing finely particulate solid polysulfide polymers that are readily dispersible in water to form a reasonably stable latex has proven to be relatively difficult. In the conventional processes for producing polysulfide polymers such as, for example, those disclosed in U.S. Patents 2,195,380; 2,278,127; and 2,363,614, an organic dihalide is added to an aqueous solution of an alkaline polysulfide containing a dispersing agent such as magnesium hydroxide. The polymer forms as the disperse phase of a latex and hence is present in particulate form. However, the particulate polymers as thus produced have a strong tendency to agglomerate, and their particle size is not readily controllable.

To some extent this difficulty can be overcome by using a dusting powder to prevent agglomeration of the particles as disclosed in U.S. Patent 2,390,853. However, an apparently insurmountable problem is presented by the fact that the polymer particles formed in this conventional process, irrespective of the method by which they are recovered from the latex in which they are formed, cannot be redispersed in water to form a new latex of acceptable stability.

One effort to circumvent this problem is disclosed in U.S. Patent 2,925,362. An organic halide of the type commonly used in making polysulfide polymers is reacted with an aqueous solution of an alkali metal thiosulfate to form an organic thiosulfate known as a Bunte salt. The water is evaporated from the reaction product to form a dry material which can be readily pulverized. The resulting powder containing the Bunte salt is then mixed with a finely divided inorganic polysulfide to form a solid concentrated composition that can be readily stored and shipped. At the point of use the composition can be mixed with water, whereupon its components react to form a polysulfide polymer as the disperse phase of a latex.

While such dry mixtures of polysulfide polymer-generating ingredients constitute an important advance, they are subject to certain limitations. The two components of the composition have somewhat different specific gravities and different particle sizes. There is thus a tendency for the ingredients to become partially segregated in their container during loading, shipment and unloading of the container. When such "demixing" occurs in transit, material removed from any particular part of the container at the point of use does not contain the proper relative proportions of Bunte salt and polysulfide reagent. One sample removed from the container may have an excess of Bunte salt and another sample an excess of polysulfide reagent.

Such mixtures are also quite sensitive to the presence of water. Minute amounts of water in either liquid or vapor form leaking into the containers in which the mixture is packaged cause a partial reaction between the ingredients and caking of the powder mixture within the container. The caked material cannot be dispersed in water to form a latex and hence is unusable for its intended purpose.

A process developed by M.M. Swaab and B.J. Sutker (U.S. Patent 3,205,203; September 7, 1965; assigned to Thiokol Chemical Corporation) is based on the discovery that when a Bunte salt is reacted in aqueous solution in the presence of a dispersing agent with an inorganic alkaline sulfide to form a polysulfide polymer latex and the freshly prepared latex is spray dried, the polymer can be recovered in the form of a fine powder which, even after an extended storage period, is readily redispersible in water to form a stable latex. Polysulfide polymer-containing products can be prepared which have an average particle size of the order of 10 to 100 microns and which can be stored as long as 12 months without losing their ability to be dispersed in water to form a stable latex.

Example: An aqueous solution of a Bunte salt was prepared by reacting 71.5 pounds of dichloroethylether (ClH_2C_2—O—C_2H_2Cl) with 310 pounds of sodium thiosulfate ($Na_2S_2O_3$) in a solution containing 30 pounds of the sodium salt of polymerized alkylnaphthalene sulfonic acid (Darvan #1) and 2.12 pounds of soda ash (Na_2CO_3) in 393 pounds of water. This solution contained the equivalent of 0.75 mol or 256.5 grams of Bunte salt per liter.

12.55 liters of this aqueous solution of Bunte salt were charged into a 10 gallon stainless steel reactor which was fitted with an agitator. To this was fed 5.45 liters of an aqueous lime sulfur solution containing 2.1 kg. of lime sulfur (CaS_4 to 5) in the following manner: 0.6 liter per minute for the first six minutes, and 0.3 liter per minute thereafter. The Bunte salt and alkaline sulfide reacted to form a polysulfide polymer in the form of a finely particulate latex having a pH of about 6.9. To prevent coagulation of the latex, the reaction mixture was agitated continuously throughout the reaction period and thereafter until the latex was fed to a dryer.

Drying of the latex was effected in the laboratory model spray dryer manufactured by Bowen Engineering, Inc., using Bowen's SS #5 atomizer nozzle. The latex was fed into a glass separatory feed funnel from which it flowed by gravity to the atomizer nozzle. The latex feed rate was 280 to 350 ml./min. and the total amount of latex fed was 2,840 ml.

The atomized latex was sprayed into a stream of air at 500°F. and 100 psig flowing at a rate of 12 lbs./hr., and the solid particles formed were separated from the air stream in a cyclone separator. A total of 1,410 grams of free flowing, off white powder was collected having a particle size within the range 10 to 100 microns and a moisture content of 3.5%.

The powder as thus prepared was tested for dispersibility by adding 3 grams thereof to 150 ml.

of water, stirring for 90 seconds and observing the extent to which it formed a reconstituted latex. It was found that the particulate product as produced above readily dispersed to form a latex, even after storage in a high humidity atmosphere for nine months at room temperature or 140°F. for one month. The particles remained suspended in discrete form without agitation for periods of about 30 minutes, after which they began to settle to the bottom of the container in a nonagglomerated state. The particles could be redispersed by merely swirling the container, even after standing in a settled condition for extended periods of time.

CELLULAR FORM

Sponge rubber is ordinarily made by incorporating a blowing agent into a very soft rubber stock so that during the heating which occurs in the vulcanization step, gas is evolved to expand the rubber into a spongy mass. Typical blowing agents are ammonium carbonate and a mixture of stearic acid and sodium bicarbonate. Sponge or cellular rubber is also made by beating air into a rubber latex to which has been added a frothing agent such as castor oil soap. The treated latex may contain a delayed action coagulant which coagulates the foam mass after it has been poured into molds and is heated by immersion in hot water. The foamed product may then be removed and dried with warm air and is ready for use. Such processes involve a number of steps and are quite time consuming. Furthermore, it is difficult to introduce sponge rubber into irregularly shaped spaces.

Use of Arylene Diisocyanates

A process developed by A. Mitchell, 3rd (U.S. Patent 2,814,600; November 26, 1957; assigned to E.I. du Pont de Nemours and Company) gives cellular materials by reacting water with a fluid polythiourethane product formed by the reaction of an arylene diisocyanate and a liquid polysulfide polymer having terminal sulfhydryl groups and having a molecular weight of from about 1,000 to 4,000. The proportions of these reactants are such that the ratio of sulfhydryl groups present in the polysulfide polymer to the isocyanate groups present in the arylene diisocyanate is from 0.1:2 to 0.4:2 and the amount of water used is from about one to about two times the amount equivalent to the unreacted isocyanate groups present in the said polythiourethane product.

The process is quite simple. The arylene diisocyanate and the liquid polysulfide polymer are first reacted together to form a stable liquid condensation product. This polythiourethane condensation product is made by heating the two reactants together in suitable proportions until reaction is complete. Heating at 100°C. for about one hour accomplishes this objective, although other temperatures and reaction times are also satisfactory. The resulting fluid polythiourethane product is cooled and may be stored until the foam is to be prepared. The viscosity of this intermediate product is ordinarily from about 750 to 75,000 centipoises at 30°C. At these viscosities the carbon dioxide which is evolved during foam formation does not escape from the mass to any extent, and still the product is capable of being properly blown into a cellular mass. If the viscosity of the polythiourethane product is too low, the carbon dioxide merely bubbles out and escapes, while if the viscosity is too high, satisfactory mixing cannot be obtained and nonuniform cellular products of higher density result.

The intermediate polythiourethane product is converted to the ultimate cellular rubberlike material simply by mixing it with water, ordinarily in the presence of a tertiary amine catalyst. The reaction mass immediately begins to foam due to the reaction of the unreacted isocyanate groups in the polythiourethane product with the water to form carbon dioxide and substituted ureas. The mixture expands to from five to fifty times its original volume and if in a confined space, the foam will fill up the space in a few minutes and within a relatively short time will cure at room temperature to yield a flexible, low density cellular material which closely resembles foamed rubber.

Example 1: To a glass flask fitted with an agitator are added 162.4 parts of Thiokol LP-2, a liquid polysulfide polymer having a molecular weight of about 4,000, having terminal sulfhydryl groups, and being cross-linked to the extent of 2%, i.e., on the average, 2% of the recurring units linked together by sulfide linkages contains either cross-linkages or sulfhydryl groups capable of forming cross-linkages. To this liquid polysulfide polymer are then added 37.6 parts of 2,4-tolylene diisocyanate and the mixture is heated to 100°C. Heating is continued at this temperature for an hour and the mixture is allowed to cool to room temperature, giving a rather viscous liquid polythiourethane product.

To 50 parts of the polythiourethane product is added a mixture of 0.79 part of water and 0.5 part of diethylcyclohexylamine. The ingredients are mixed quickly and thoroughly and allowed to stand. Foaming begins at once. Within about two minutes the mass has foamed to its maximum volume and in an hour has set completely. The result is a flexible foam having good cell structure and resilience and having the general appearance of sponge rubber. Its density is 5.74 lbs. per cubic foot.

Example 2: Using the procedure described in Example 1, foams are prepared using different Thiokol liquid polysulfide polymers and varying the proportions of the ingredients, as follows:

	Thiokol Liquid Polymer			Ratio of sulfhydryl to isocyanate groups	Water, % of theory used
	Type	MW	Percent cross-linking		
A	LP-2	4,000	2	0.25/2	100
B	LP-2	4,000	2	0.125/2	100
C	LP-32	4,000	0.5	0.375/2	100
D	LP-32	4,000	0.5	0.25/2	100
E	ZL-152	4,000	0	0.375/2	100
F	ZL-152	4,000	0	0.25/2	100
G	ZL-137	1,000	6	0.375/2	100
H	ZL-137	1,000	6	0.375/2	120
I	LP-3	1,000	2	0.375/2	120

The properties of the resulting foams are shown in the table on the following page.

	Cell Structure	Foam Character	Density, lbs./cu. ft.
A	Good	Flexible and resilient	3.00
B	Fair	Flexible and resilient	2.90
C	Good	Soft and flexible	--
D	Good	Soft and flexible	5.45
E	Good	Flexible and resilient	5.44
F	Fair	Flexible and resilient	3.71
G	Fair	Flexible and resilient	1.33
H	Good	Flexible and resilient	1.55
I	Fair	Flexible and resilient	1.65

Use of Polyester-Polyisocyanate

The products of a process developed by E. Simon and F.W. Thomas (U.S. Patent 2,929,794; March 22, 1960; assigned to Lockheed Aircraft Corporation) fall into two general categories or classes: resins A and resins B. The components of a type A resin or a type B resin are not all permitted to compete one with the other simultaneously for the active or functional groups of a coreactant molecule. In the preparation of the resins it is preferred that an initial reaction occur in accordance with a given order of component addition to obtain reproducible, controllable and predetermined results.

A preferred order of reaction of the components in preparing an A type resin is: first, a reaction between (1) a diisocyanate, polyisocyanate, polyisothiocyanate, or blends thereof, and (2) diols, such as polyglycols, linear aliphatic glycols, or substituted or unsubstituted modifications and blends thereof. The diols may be replaced to an extent up to 40% by weight of nitrogen base resinous compositions having a molecular weight of between 500 and 10,000, these nitrogen base resinous compositions being the reaction products of dibasic alkyl carboxylic acids with amino alcohols and alkylamines, and the self esterified reaction products of linear alkyl amino acids, or by polymeric polysulfides within the molecular weight range of from 200 to 10,000, the latter being known commercially as Thiokol base compounds. These Thiokol compounds are preferably the reaction products of dihaloalkyl-formals, sodium polysulfide and trihaloalkyls such as reaction products of dichloroethyl-formal, sodium polysulfide and trichloropropane. The resins, which include the Thiokols as components, are particularly well suited for use as or for preparing sealant materials, putties, and the like.

The second stage in the preparation of an A type resin is the addition to the initial reaction product of a bifunctional acid such as a linear hydroxy acid, a dicarboxylic acid, an amino acid, or unsaturated, substituted modifications or blends of the same and water, the resultant mixture reacting to form what may be termed an intermediate. The third stage in the preparation or reaction of an A type resin is the reaction between the prereticulated polymers resulting from the first and second stages of reaction above described and a reticulating agent which is a trifunctional molecule containing alcohol, carboxylic acid, amine or mercaptan groups, or mixtures of the same. The preparation or reaction of an A resin is concluded when the resultant resin reaches an amine equivalent of from 150 to 1,000, an amine equivalent value of approximately 500 being preferred. In describing the preparation or reaction of the A resins as well as the B resins the reticulating agents referred to are comprised of a molecule containing more than two functional groups comprising isocyanates, isothiocyanates and the above mentioned polylabile hydrogen compounds.

The preferred method of preparing or reacting a B type resin first includes the bringing about of a reaction between one or more diisocyanates and one or more diols, this being the same as the first stage reaction in preparing an A resin. The second stage in preparing a B resin involves the reaction between the product of the first reaction and polymers or blends of polymers, which will be designated C class resins, and water and, thirdly, and finally, the reaction between the prereticulated polymer resulting from the first and second reactions and a reticulating agent, namely a trifunctional molecule containing alcohol, carboxylic acid, amine or mercaptan groups or mixtures of these. The reaction in preparing a B class resin is concluded when the amine equivalent of the resin reaches a value of from 150 to 1,000, the preferred value being approximately 350. The resins of the C class utilized in preparing a B type resin, as just described, may be groups or classified, as follows: C Type I, polyesters such as the reaction products between dibasic acids and dihydric alcohols; C Type II, nitrogen base polymers such as polyamides, and/or esteramides; C Type III, silicon base polymers such as silico polyesters; C Type IV, sulfur base polymers, such as polymeric polysulfides.

In synthesizing or preparing the Class B resins, modifying resins are used as reactant intermediate resins, these intermediate resins containing or having labile hydrogen atoms capable of reacting with the isocyanate or polyisothiocyanate-isocyanate blends. These modifying resins are polyesters, polyamides and esteramides, silicol polyesters or polysulfide resins. The polyesters are the reaction products of saturated, unsaturated, substituted or unsubstituted alkyl polyfunctional alcohols and acids, preferably dibasic carboxylic acids and alcohols, and have an acid number range of from 0.1 to 200, preferably about 50. The polyamides and esteramides are the self polymerization reaction products by way of the bifunctional condensation of amino acids, of the bifunctional reaction products of amino alcohols with dibasic carboxylic acids or mixtures thereof. These polyamides and esteramides preferably have a molecular weight of from 500 to 10,000. The silico polyesters useful as the modifying resins are the reaction products of organic bifunctional silanols, silacols or condensed silacols, such as linear alkyl polyhydroxy polysiloxanes with the alcohols or acids or amines such as used in the preparation of the above described polyesters and polyamides. The molecular weight of the silico polyesters may range between 500 and 10,000.

The following is a reaction chart for preparing a typical A resin. The synthesizing of the resin should be carried out in a glass lined reactor, a stainless steel reactor, or the like, and the temperatures, as well as the reaction times should be controlled. The time sequences will, of course, vary with the size of the batch or run and the tailoring of the final resin to the desired or specified amine equivalent. The following chart covers the preparation of a pilot run in which approximately one gallon of the resin is produced, it being understood that the components referred to in the chart are merely illustrative of this type of resin. The chart is shown on the following page.

Physical Form

Phase I

Total Elapsed Time	Temperature, °F.	Amine Equivalent	Order of Component Addition to Reactor
0	70	–	Isocyanate added.
0	70	–	Diol added.
20	150	–	Diol addition complete.
40	150	–	Polyurethane reaction complete.

Phase II

40	150	–	Difunctional acid added.
90	150	–	Water added.
100	150	–	Reaction finished.

Phase III

100	150	–	Reticulating agent added.
105	150	–	Reticulating agent addition complete.
135	150	–	
145	250	–	
165	250	–	Reaction finished.
200	70	550	(Cooling period.)

The following chart or schedule illustrates the preferred sequence of adding or incorporating typical components in preparing a type B resin relative to the temperatures, timing, etc. The above comments concerning the preparation of the A resins are applicable to the synthesizing of the B resins.

Phase I

Time	Temp., °F.	Component Addition
0	70	Meta-toluene diisocyanate.
0	70	Polypropylene glycol water mixture.
20	200	Polypropylene glycol addition finished.
50	200	Reaction time.

Phase II

90	200	A Class C Type I-a polyester resin added.

Phase III

100	200	1,2,6-hexane triol added.
110	260	1,2,6-hexane triol all in.
125	260	Final reaction amine equivalent 308.

Use of Hydrazine-Type Compounds

In a process developed by B.A. Hunter (U.S. Patent 3,114,723; December 17, 1963; assigned to United States Rubber Company), the expansion of liquid polysulfide rubbers is achieved without necessity for the application of heat by incorporating a hydrazine compound into a liquid-type polysulfide rubber composition and treating the composition with an oxidizing curing agent conventionally used in the curing of such liquid polysulfide rubbers. The oxidizing curing agent performs the double function of concomitantly curing the rubber and oxidatively decomposing the hydrazine blowing agent to produce nitrogen gas as an expanding agent in the rubber. Little or no heat is necessary to bring about the production of gaseous products from the blowing agents utilized in the process. The formation of gas in the process is achieved by the oxidative decomposition of the hydrazine compound by the action of the oxidizing curatives employed in the curing of the liquid polysulfide rubber composition.

Example: The polysulfide rubber employed in this example was a condensation product of 98 mol percent bis-beta-chloroethyl formal and 2 mol percent of 1,2,3-trichloropropane with sodium polysulfide, having an equivalent weight of 1,710 by mercaptan end group analysis, and having a viscosity of 560 poises at 30°C. as measured by means of a Brookfield viscometer. Such a liquid material is commercially available under the name Thiokol LP-2. 100 parts of the foregoing polymer was mixed with 30 parts of semireinforcing carbon black (Pelletex) and 1 part of stearic acid.

The resulting mixture is known as Thiokol T-13-A liquid polymer. To this was added 0.44 gram of hydrazinium formate. The materials were thoroughly mixed, and then 5.0 g. of a paste consisting of 50 parts of lead dioxide, 45 parts of dibutyl phthalate, and 5 parts of stearic acid (Thiokol Accelerator C-5) was stirred into the mix. The thoroughly blended mixture was allowed to stand overnight at room temperature. Next day the product was found to be cured to a well expanded rubber exhibiting a fine, uniform cellular structure. The density of the product was found to be 0.79. A similarly prepared rubber omitting the hydrazinium formate exhibited no cellular structure, and the measured density was 1.41.

Polysulfide Polyurethanes from ω,ω'-Dihydroxydialkyl Polysulfide

It is known according to U.S. Patent 2,814,600 (cited previously) to prepare cellular resins from arylene diisocyanates and liquid polysulfide polymers. For example, one is made by heating sodium polysulfide with a mixture containing 98% di(chloroethyl)formal and 2% of 1,2,3-trichloropropane to form a polymer of fairly high molecular weight and then reducing this polymer to split some of the polysulfide linkages and convert them to sulfhydryl groups. The cellular resin prepared by this process has mechanical properties inferior to those of conventional cellular polyurethane resins and, moreover, it is difficult to carry out the reaction because of the unreactive nature of the —SH group thus requiring long reaction times, high reaction temperatures and/or extremely active catalysts.

Another U.S. Patent 2,929,794 (also previously cited) proposes the use of sulfur base polymers such as polymeric polysulfides for reaction with prepolymers prepared from a polyisocyanate and a diol. No method is proposed in the patent for introducing terminal groups other than sulfhydryl groups so that the polysulfide will react with an organic polyisocyanate under ordinary reaction conditions. The polythiourethanes based on these polymeric poly-sulfides have an extremely unpleasant odor in addition to having very poor mechanical properties. However, their resistance to organic solvents and particularly aromatic solvents is outstanding. Polyurethanes based on conventional polyesters, polyethers, polythioethers and the like are not as resistant as desirable to aromatic solvents but have excellent mechanical properties. Furthermore, the polysulfides are known to split at relatively low temperatures at the polysulfide linkages to form lower molecular weight products containing terminal sulfhydryl groups.

A process developed by G. Dankert and H. Holtschmidt (U.S. Patent 3,169,119; February 9, 1965; assigned to Farbenfabriken Bayer Aktiengesellschaft, Germany) reacts an organic polyisocyanate with a polyhydroxyl compound having a molecular weight of from about 500 to 10,000 which has been prepared by a process which comprises reacting an ω,ω'-dihydroxy-dialkyl polysulfide with formaldehyde. Included are polyurethanes which have extremely low sensitivity to aromatic solvents combined with good tensile strength, good breaking elongation, good elasticity and which only have a very slight odor.

These polyurethanes prepared from sulfur base polymers are preferably prepared from poly-acetals which have terminal aliphatic hydroxyl groups. The polyhydroxyl compound used for reaction with the organic polyisocyanate is prepared by reacting an ω,ω'-dihydroxydi-alkyl polysulfide which preferably has the general formula $HO(R-S_x)_yROH$ wherein R is an alkylene radical preferably having from 2 to 5 carbon atoms, x is a number of from 2 to 5 and y is an integer preferably below about 15 with a formaldehyde yielding substance preferably in the presence of a catalyst which will split off water to form a product having a molecular weight of from about 500 to 10,000.

Example 1: About 1,540 parts of β,β'-dihydroxydiethyl disulfide and about 270 parts of paraformaldehyde are heated with about 300 cubic centimeters of benzene and while stirring to about 80°C. Sulfur dioxide is introduced into this mixture until a pH value of about 2 to 3 is reached. After completing azeotropic dehydration, the sulfurous acid which is present is neutralized with gaseous ammonia and a polyacetal is obtained which has an —OH number of about 59.

About 100 parts of this polyacetal are dehydrated for about half an hour at about 130°C. and 18 mm. and stirred with about 32 parts of 1,5-naphthylene diisocyanate and about 7 parts of butane-1,4-diol at about 130°C. The viscous melt is cast on to a support and finally heated for about 24 hours at about 100°C.

Thickness of the test plate	4.9 mm.
Tensile strength	135 kg./cm.^{2}
Swelling in benzene after 7 days	10.2%

Example 2: About 5,580 parts of β,β'-dihydroxydiethyl trisulfide and about 800 parts of paraformaldehyde are heated with addition of about 18 parts of p-toluene sulfonic acid while stirring well and introducing nitrogen to about 110°C. After splitting off about 250 cubic centimeters of water, the substance is further condensed under partial vacuum. After neutralization with potassium carbonate, a polycondensate is obtained which has an hydroxyl

number of about 118. About 100 parts of this polyacetal are dehydrated for about 30 minutes at about 130°C./18 mm. and stirred with about 35 parts of 1,5-naphthylene diisocyanate and about 2.4 parts of butane-1,4-diol at about 130°C. The viscous melt is poured on to a support and finally heated for about 36 hours at about 100°C.

Thickness of the test plate	4.2 mm.
Tensile strength	165 kg./cm.^{2}
Swelling in benzene after 7 days	3.8%

Example 3: About 100 parts of the polyacetal obtained according to Example 1 and having an —OH number of about 59 are stirred with about 17 parts of 80% of 2,4- and 20% 2,6-toluylene diisocyanate and about 0.9 part of water for about 45 minutes at about 60°C. The viscous melt is poured on to a support and is then finally heated for about 12 hours at about 100°C. Using a rubber roller, about 20 parts of carbon black, about 0.5 part of stearic acid and about 8 parts of dimeric 2,4-toluylene diisocyanate are incorporated into this storable material by rolling. The sheet is vulcanized in a press for about 60 minutes at about 140°C.

SINGLE-PACKAGE COMPOSITIONS

CASTABLE LIQUID POLYMER AND ITS CURE

A process developed by G.M. Le Fave and R.J. Wyman (U.S. Patent 3,215,677; November 2, 1965; assigned to Coast Pro-Seal & Mfg. Co.) is fundamentally a system for curing a castable liquid polymer, said system being composed of the premixed composition to be cured and a molecular sieve loaded with a chemical which initiates the cure. At the proper time the initiating chemical is expelled from the sieve into contact with the composition to be cured, thereby effecting the cure. Molecular sieves, as used in this process, are described in "Chemical Loaded Molecular Sieves" (Form No. F-1311), published by Linde Company, Division of Union Carbide Corporation (1959); in "Chemical Loaded Molecular Sieves in Rubber and Plastic" (Form No. F-1349), published by Linde Company, Division of Union Carbide Corporation (1959); and in U.S. Patent 3,036,980.

The fundamental reactants are those described in U.S. Patent 3,138,573 (see Chapter 3). In general, the acceptor is a sulfone activated diethylenic compound. The donor is a branch chain liquid polymer terminated with thiol groups and being polyfunctional with respect to the thiol groups, the functionality being greater than 2 and less than 6. The body of the liquid polymer is composed of repeating units of alkylene groups of 1 to 10 carbons which may, if desired, be substituted with halogen, or arylene groups of 1 to 10 carbons which likewise may, if desired, be substituted with halogen. The groups are connected by linkages of ester, monosulfide, polysulfide, urethane, oxide, or combinations thereof. The process may be expressed symbolically in the following figures:

$$H_2C{=}CHSO_2{\sim}\!\!\sim\!\!R{\sim}\!\!\sim SO_2CH{=}CH_2 + HS{\sim}\!\!\sim R^1{\sim}\!\!\sim SH$$

liquid prepolymer (+ Residual catalyst)   Polysulfide liquid polymer

$$\Big[H_2C{-}CH_2SO_2{\sim}\!\!\sim R{\sim}\!\!\sim SO_2CH_2{-}CH_2{-}S{\sim}\!\!\sim R^1{\sim}\!\!\sim S \Big]$$

Solid polymer

In the above illustration the prepolymer and polysulfide liquid polymer mixture are unstable and cannot be kept in the same container under normal temperature for any appreciable length of time. This is because the system is sufficiently basic by virtue of the residual catalyst and further polymerization continues. Polymerization can be effectively inhibited to make a completely stable compound by using acids or anhydrides of 6 to 16 carbons, e.g., pyromellitic dianhydride. This effectively neutralizes the residual catalyst and makes the system slightly acidic, to prevent the polymerization reaction. This may be shown schematically as follows:

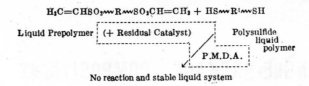

$$H_2C=CHSO_2\text{w}R\text{w}SO_2CH=CH_2 + HS\text{w}R^1\text{w}SH$$

Liquid Prepolymer (+ Residual Catalyst) | Polysulfide liquid polymer

P.M.D.A.

No reaction and stable liquid system

With the mixture charted above, is mixed the molecular sieve loaded with a base which serves as a latent catalyst. Once the base catalyst is loaded into the sieve it is held sufficiently strongly within the structure to isolate it from the environment. The base catalyst remains isolated and confined within the sieve until it is released by heat or by displacement by another chemical to which the molecular sieve has a greater attraction.

The molecular sieves have an extremely high preferential absorption for water, so that even a small amount of moisture in the air is sufficient to drive the base out of the sieve and into neutralizing reaction with the inhibitors. The action of the expelled base is essentially that of a catalyst since it does not become consumed as the polymer cures progressively. When this system is stored in a sealed container at room temperature, the stability is about 6 months. Upon exposure to atmospheric moisture the amine is released from the molecular sieve by the preferential adsorption and it proceeds to neutralize the acidic stabilizer and catalyze the polymerization reaction by making the total system slightly basic.

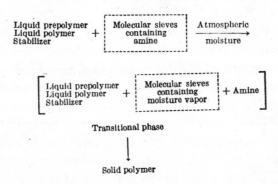

The following formulations illustrate the use of a variety of stabilizers and chemically loaded sieves. These formulas include adhesion promoters and thixotropic agents, used where the end product is to be a non-sag sealant or caulking compound.

Polysulfide Liquid Polymer[1]	400	400	400	400	400
Liquid Prepolymer[2]	500	500	500	500	500
Inert Fillers	525	525	525	525	525
Thixotropic Agent[3]	6	16	16	16	16
Adhesion Promoter[4]	25	25	25	25	25
Dessicant[5]	25	30	30	30	30
Stabilizer:					
Pyromellitic Dianhydride	0.8	0.8	0.8	0.8	0.8
Adipic Acid	—	—	—	—	—
Maleic Anhydride	—	—	—	—	—

(continued)

Phthalic Anhydride	—	—	—	—	—
Curing Agent, Loaded Molecular Sieves:					
15% Triethylene Diamine	15	—	—	—	—
15% Triethylamine	—	23	—	—	—
15% Trimethylamine	—	—	15	—	—
15% N, N, N', N'-Tetramethyl–1, 3–Butanediamine	—	—	—	14	—
15% N, N, N', N'-Tetramethyl Ethylene Diamine	—	—	—	—	14
15% 1, 2, 4-Trimethylpiperazine	—	—	—	—	—
15% N-Methylmorpholine	—	—	—	—	—

[1] Liquid Polysulfide Resin of 1,000 molecular weight, Thiokol LP-33.

[2] A Thiokol LP-3-D.V.S. prepolymer with an equivalent weight, based on unreacted sulfone activated ethylenic groups, of about 800.

[3] Stearic acid, as an example.

[4] Phenolic resin, as an example.

[5] Molecular Sieve powder, Type 5-A, Linde Company, Division of Union Carbide Corp.

Polysulfide Liquid Polymer[1]	400	400	400	400	400
Liquid Prepolymer[2]	500	500	500	500	500
Inert Fillers	525	525	525	525	525
Thixotropic Agent[3]	16	16	16	16	16
Adhesion Promoter[4]	25	25	25	25	25
Dessicant[5]	30	30	30	30	30
Stabilizer:					
Pyromellitic Dianhydride	0.8	0.8	—	—	—
Adipic Acid	—	—	1.0	—	—
Maleic Anhydride	—	—	—	1.0	—
Phthalic Anhydride	—	—	—	—	1.0
Curing Agent, Loaded Molecular Sieves:					
15% Triethylene Diamine	—	—	9.0	9.0	9.0
15% Triethylamine	—	—	—	—	—
15% Trimethylamine	—	—	—	—	—
15% N, N, N', N'-Tetramethyl–1, 3 –Butanediamine	—	—	—	—	—
15% N, N, N', N'-Tetramethyl Ethylene Diamine	—	—	—	—	—
15% 1, 2, 4-Trimethylpiperazine	14	—	—	—	—
15% N-Methylmorpholine	—	20	—	—	—

[1] Liquid Polysulfide Resin of 1,000 molecular weight, Thiokol LP-33.

[2] A Thiokol LP-3-D.V.S. prepolymer with an equivalent weight, based on unreacted sulfone activated ethylenic groups, of about 800.

(continued)

3 Stearic acid, as an example.
4 Phenolic resin, as an example.
5 Molecular Sieve powder, Type 5-A, Linde Company, Division
    of Union Carbide Corp. The Thiokols are described in "Poly-
    sulfide Polymers (Thiokol)", by J.S. Jorczak, contained in the
    larger work "Introduction to Rubber Technology", by Maurice
    Martin, at page 378-80, Reinhold Publishing Corporation, New
    York, 1959. The molecular sieve powder is described in the
    aforementioned publications of Linde Company.

ACTIVATION BY MOISTURE

A method developed by I.P. Seegman, L. Morris and P.A. Mallard (U.S. Patent 3,225,017; December 21, 1965; assigned to Products Research Company) involves a one-part, stable liquid polysulfide polymer composition and method which can be completely cured without agitation immediately prior to deposition. The composition may be deposited in place and then cured solely by contact of its surface with surroundings containing essentially only moisture. Thus, the method eliminates any mixing step immediately prior to use and consequently eliminates air bubbles in the cured polysulfide rubber. Also, the composition may be packaged in a single container and applied directly to the place where it is used. After deposition in place, merely by contact with atmospheric air, even thick bodies of composition may be cured without the addition of separate curing agents.

In general, the process involves a one-part, stable, hygroscopic liquid polymer composition comprising essentially a liquid polyalkylene polysulfide polymer. The polymer has thoroughly dispersed therein a dormant curing agent for it which is activated by the presence of moisture. Likewise, the polymer has thoroughly dispersed therein a water-soluble deliquescent accelerating agent adapted to attract and absorb moisture from the surroundings and to hasten the curing of the polymer by the curing agent. The polymer may be intially dried to remove moisture or, preferably, the deliquescent accelerating agent may also be a desiccating agent to dry the polymer. Alternatively, the polymer may have thoroughly dispersed therein a single desiccating, deliquescent, dormant curing and accelerating agent which is adapted to dry the polymer, to attract and absorb moisture from the surroundings, to cure the polymer when activated by the presence of moisture and to hasten the curing of said polymer. Such surroundings may include a body of water or a body of gas containing essentially only moisture such as atmospheric air of normal humidity.

Example 1: To polyalkylene polysulfide having a molecular weight of about 4,000, a viscosity of about 40,000 centipoises and 0.5% cross-linking, i.e., Thiokol LP-32, are added the following components in the proportions by weight indicated:

Thiokol LP-32	100
Titanium dioxide	18
Calcium carbonate	45
Plasticizer[1]	15

(continued)

Adhesive resin[2]   5
Calcium peroxide (60%)   10
Barium oxide   10

[1] 47 parts of Arochlor 1254, a chlorinated diphenyl produced by Hercules Powder Co. and 10 parts of HB-40, a hydrogenated terphenyl by Monsanto Chemical Co.
[2] 80% solution of Epon 1001, an epoxy resin, in methyl ethyl ketone produced by Shell Chemical Co.

All ingredients were thoroughly mixed in a paint mill and the resulting formulation was loaded into sealed cartridges suitable for use in a standard extrusion gun. At intervals, as desired, some of the cartridges extruded onto test panels. The material cured into tough rubberlike compositions which, when tested, showed physical and chemical properties at least equivalent to similar compositions cured in the usual manner by mixing.

Properties

Stability at 120°F.	30 days.
Stability at 75°F.	6 months.
Tack-free time	24 hours.
Cure time at 75°F. and 50% relative humidity	30 days for 1/4 in. thickness.
Cure time at 120°F. and 100% relative humidity	2 days for 1/4 in. thickness.

Adhesion (after immersion 5 days in water;
     MIL-S-7502 Peel Test):

	(lbs./in.^{2})
Aluminum Alloy 6061	35
Glass	30
Stainless steel	35
Galvanized steel	32

Examples 2 through 5: Some examples of other compositions in parts by weight which may be used are as follows:

		2	3	4	5
Thiokol LP-32		100	100	100	100
Titanium Dioxide		20	20	20	20
Calcium Carbonate		40	40	40	40
Plasticizer (1)		10	10	10	10
Adhesive Resin (2)		5	5	5	5
Zinc Peroxide	(4)	10	10	------	------
Plasticizer (3)		10	10	------	------
Calcium Peroxide	(4)	------	------	3	10
Plasticizer (1)		------	------	6	20
Sodium pyrophosphate		------	------	------	------
Peroxide	(4)	------	------	2	0.5
Plasticizer (5)		------	------	2	0.5
Barium Oxide		------	10	10	10
Plasticizer (5)	(4)	------	10	3	10
Toluene		------	------	3	------
Sodium Hydroxide		0.5	0.25	------	------
Plasticizer (5)	(4)	1	0.5	------	------
Toluene		0.5	0.25	------	------
Toluene (6)		3.75	3.75	3.75	3.75

(1) Arochlor 1254.
(2) 80% solution of Epon 1001 in methyl ethyl ketone.
(3) Arochlor 1242.
(4) Dispersion.
(5) HB-40.
(6) Dispersion prepared with 10 parts of the Thiokol LP-32.

STABILITY DURING STORAGE VIA CHELATING AGENTS

An earlier method covered by U.S. Patent 2,940,958 (Chapter 3) involves mixing a polymer with a water-activatable curing agent and a hydrated salt. So long as the composition is kept at low temperature, it remains stable. When the composition is heated, the water of hydration of the hydrated salt is released and activates the curing agent to cause the composition to cure to an elastomeric form.

While such a one-package polysulfide sealant composition is satisfactory for some applications, it is also subject to certain undesirable limitations. For example, in many cases it is desired to use the sealant composition in environments which cannot readily be heated to bring about a cure of the sealant composition. Also many of the hydrous salts begin to release water of hydration at about 100°F., and then compositions of this type become somewhat unstable in southern latitudes or in temperature latitudes at midsummer temperatures. Since the polysulfide polymers are rather viscous, it is convenient to carry out the compounding of the compositions at a temperature somewhat above room temperature. However, an elevated temperature mixing procedure cannot very well be used when hydrated salts are employed as a component of the mixture.

A further disadvantage of the compositions containing hydrated salts is that they impose serious limitations on cure rate. If it is necessary to slow down the cure rate as required for certain caulking applications, this can be accomplished by reducing the amount of hydrated salt present. However, such a reduction in the amount of hydrated salt results in a product with relatively poor physical properties.

A process developed by J.S. Jorczak and A.W. Volk (U.S. Patent 3,247,138; April 19, 1966; assigned to Thiokol Chemical Corporation) produces a one-package polysulfide coating and sealing composition which remains stable when stored at atmospheric temperatures for an indefinite period of time, and which, when used for coating or sealing applications after an extended storage period, readily cures to provide a rubbery seal or coating having good physical and chemical properties. Also, the product has a curing agent which remains inert when the composition is in a sealed container but is activated on exposure to either atmospheric oxygen or water to produce a cure of the polysulfide polymer.

The curing catalyst comprises two principal components, namely, a metal soap and a chelating agent. Both the metal soap and chelating agent are employed in solution in an inert organic solvent to facilitate their dispersion in the polymer composition. The metal soaps comprise those that have previously been employed as driers in paint mixtures and include the naphthenates, octoates, and tallates of cobalt, manganese, iron and lead and mixtures of such metal soaps. Any of the various commercially available solvent solutions of these soaps, which commonly contain from 5 to 25% by weight of metal, may be used satisfactorily. The amount of metal soap used in these compositions varies as a function of the curing rate desired. Usually the quantity of soap is such as to provide from 0.1 to 0.6% by weight of metal based on the weight of the polymer. The gross weight of the soap incorporated in the composition is usually from about 0.2 to 4% by weight of the polymer.

Certain chelating agents previously proposed as deactivators in paint and petroleum compositions have been found especially effective in these compositions. For example, it has been found that a commercial chelating agent sold under the trade name ACTIV-8 and comprising 38% by weight of 1,10-phenanthroline, 52% of normal butyl alcohol and 10% of 2-ethylhexoic acid, when incorporated in a polysulfide composition with a metal soap as described herein, produces a slow-curing, fast-skinning composition useful for many sealant applications. Another chelating agent that produces acceptable curing activity when compounded with a metal soap in the compositions is sold under the trade designation MDA and consists essentially of 80% N,N'-disalicylidene-1,2-diaminopropane and 20% of toluene.

Since the curing catalysts used in these compositions are activated by oxygen, it is important that compounding be carried out in an oxygen-free atmosphere and that the compositions be maintained in such an atmosphere until they are used. A typical procedure for preparing the compositions is as follows.

Mixing is effected in a Bake-Perkins sigma mixer having a 0.7 gallon capacity and equipped with connections to suitable vacuum and pressurizing sources. The polymer is placed in the mixer and a blend of the compounding ingredients other than the curing catalyst are added thereto. Mixing of the liquid polymer and subordinate compounding ingredients is continued for about 15 minutes after which the mixer is placed under a 15 to 20 inch vacuum. The curing catalyst comprising a mixture of metal soap solution and chelating agent solution is then added to the mixer. Mixing is continued for about two hours under vacuum to remove oxygen occluded in the polymer and the compounding ingredients, as well as from the atmosphere within the mixer above the mix. During this two hour period the solvent associated with the curing catalyst evaporates. At the end of the two hour period of mixing, nitrogen or other inert gas is introduced into the mixer to fill the free space above the mixture to a pressure 5 to 10 psi gauge. Mixing of the batch is terminated about a half hour after the introduction of the inert gas into the mixer.

Compositions as thus prepared can be packaged in collapsible metal tubes or rigid containers, as desired. When proper precautions are taken to exclude moisture and oxygen from the container, these compositions are characterized by outstanding long term stability. Storage temperatures may vary from -20° to +160°F. Without activating any noticeable cure. Also specimens have been stored in metal tubes at 75°±20°F. for more than a year and a half and at 120°F. for more than a year without any significant curing. In cases where the compositions are packaged in collapsible tubes, metal tubes are preferred over plastic tubes because the former are more nearly impervious to moisture and atmospheric oxygen. Such collapsible tubes provide a convenient method of applying the composition to cracks or crevices that are to be sealed therewith.

ISOCYANATE-BLOCKED POLYSULFIDE COMPOSITIONS

A process developed by E.R. Bertozzi (U.S. Patent 3,361,720; January 2, 1968; assigned to Thiokol Chemical Corporation) results in polysulfide polymer compositions wherein mercaptan-functional polysulfide polymers have their reactive mercaptan groups blocked by monofunctional organic isocyanates to form heat sensitive isocyanate-capped mercaptan group-containing organic polysulfide polymers. It also gives curable and time-stable admixtures of such capped polysulfides and curing agents.

The process thus provides curable and stable one-package compositions of mercaptan-functional polysulfide polymers and curing agents therefor, and bin-stable curable compositions of mercaptan-functional polysulfide crude rubbers in admixture with curing agents therefor.

Example 1: The following is a description of the preparation and cure of IBLP-3, an isocyanate-blocked liquid polysulfide polymer of about 1,000 molecular weight. A 2-liter, 3-necked reaction flask fitted with stirrer, condenser with drying tube open to the atmosphere, and a gas admission tube for providing a blanket of dry gas therein, was sequentially charged with agitation with 1 mol of LP-3 polymer, a liquid dimercaptan-functional polysulfide polymer of molecular weight of about 1,000, previously dried of water, having as its repeating unit $-(C_2H_4OCH_2OC_2H_4SS)-$ and having a 2% cross-link provided by trichloropropane, and with 2.2 mols of phenyl isocyanate under a blanket of dry nitrogen gas. The reaction mixture was elevated to 192° to 203°F. and maintained there for about 5 hours and then cooled to ambient to provide IBLP-3, an isocyanate-blocked liquid polysulfide polymer of this process. IBLP-3 was of similar color but slightly more viscous than LP-3 polymer.

The blocked liquid polysulfide polymer IBLP-3 was uniformly admixed with lead dioxide, a curing agent for mercaptan-functional polysulfide polymers, in quantity of 7.5 parts by weight per 100 parts by weight of polymer. The curable composition was permitted to stand at ambient temperatures of about 75°F. and at the end of at least 16 hours no perceptible signs of cure, such as thickening of the composition was observed. Admixtures of other non-blocked mercaptan-functional polysulfide liquid polymers and lead dioxide of otherwise identical constitution to this composition completely cured to form solid rubbers at about 75°F. in from 1 to at most 4 hours. The composition was then heated at about 212°F. for about 30 minutes to provide a fully cured rubbery polysulfide vulcanizate.

Example 2: The following is a description of the preparation of IBLP-2, an isocyanate-blocked liquid polysulfide polymer of about 4,000 molecular weight. In similar manner, 1 mol of LP-2 polymer, a dimercaptan-functional polysulfide polymer similar to LP-3 polymer in all ways but having a molecular weight of about 4,000, was reacted with about 2.2 mols of phenyl isocyanate to provide IBLP-2, an isocyanate-blocked liquid polysulfide that was of similar color but more viscous than LP-2 polymer. IBLP-2 in admixture with curing agents was capable of long-termed stability at ambient storage temperatures, and cure to a fully vulcanized rubber at temperatures of about 150° to 400°F. within 0.1 to 6 hours.

Examples 3 through 12: The following describes the preparation and cure of IBSTCR, an isocyanate-blocked polysulfide crude rubber. Mercaptan-functional polysulfide crude rubbers, which usually have molecular weights in excess of 100,000 to about 2 million have relatively poor bin-stability, which is to say that upon relatively short-termed storage either in admixture with curing agents therefor or even in contact with air, the oxygen of which acts as a slow curing agent therefor, the rubbers will cure to the degree that they can no longer be worked on a mill so as to permit admixture with necessary compounding ingredients, and cannot be formed to the shape of end desired vulcanized articles. In essence they lose the needed piezo-thermoplasticity to be worked prior to cure.

In the following examples, a decyl-mercaptan-functional polysulfide crude rubber designated STCR, formed with repeating units of $-(C_2H_4OCH_2OC_2H_4SS)-$ and a 2% cross-link imparted by trichloropropane, was compounded with various ingredients on a rubber mill at ambient temperatures, and observed as to its bin–stability. In Example 3, no curing agents were used. In Examples, 3, 5, 7, and 9, a small quantity of IBLP–3, prepared as above was added, which contained enough free excess phenyl isocyanate from the blocking step to substantially block all of the mercaptan groups of the STCR crude rubber. In Examples 4 to 9 sundry curing agents and mixtures were used, and are designated in the table of example as C1, which is quinone dioxime (GMF); C2, zinc oxide; C3, 2% tincture of iodine and C4, lead dioxide.

A portion of each of the milled rubber stocks was stored at ambient temperatures and observed daily for loss of workable properties, or bin–stability. Another portion of the milled stocks was cured at 287°F. for 30 minutes to be tested for vulcanizate properties. To prepare the stocks, two masterbatches were prepared on a rubber mill according to the recipes:

Masterbatch A:	Parts by Wt.
STCR	100
Stearic acid	1
Carbon black, SRF #3	60
Masterbatch B:	
STCR	97
Stearic acid	1
Carbon black, SRF #3	60
IBLP–3	3

The masterbatches were then heated for about 24 hours at about 200°F. to remove traces of moisture, and to aid in completion of the blocking of the crude rubber in Masterbatch B. The masterbatches were then cooled to ambient. The aforesaid curing agents were then milled into the masterbatches at ambient temperatures in the quantities listed in the table of ex–amples to provide curable compounded crude rubber stocks. A portion of each stock was stored at ambient temperatures and observed daily for loss of bin–stability. Another portion was vulcanized to provide cured rubbers with properties as listed below.

	Example						
	3	4	5	6	7	8	9
Recipe, in parts by weight (p.b.w.):							
Type of masterbatch [1]	B	[6] A	[6] B	A	B	A	B
Amount of masterbatch	161	161	161	161	161	161	161
Type of curing agent [1]		C1 & 2	C1 & 2	C2 & 3	C2 & 3	C4	C4
Amount of curing agent	0.0	1.5 & .5	1.5 & .5	1.5 & .5	1.5 & .5	3	3
Uncured stocks:							
Bin stability, in days	>500	66	500	17	500	3	491
Mooney viscosity, ML–4 [2] initial upon preparation	25						
Vulcanizate Properties:							
Tensile strength, in p.s.i.[3]	([5])	1,100	850	1,200	1,100	1,190	960
Ultimate elongation, in p.s.i.[3]	([5])	300	360	390	420	250	310
Hardness, in Shore A durometer degrees [4]	([5])	65	58	64	59	68	62

[1] As in text.
[2] Test method ASTM D–1646–5?T.
[3] Test method ASTM D–412–51T.
[4] Test method ASTM D–676–59T.
[5] No cure.
[6] 2 p.b.w. of stearic acid added to recipe.

USE OF ZEOLITIC MOLECULAR SIEVES

A number of moisture activated one-part polysulfide polymer compositions have been developed in U.S. Patent 3,225,017 (previously mentioned). These one-part moisture activated compositions, however, have the disadvantage of having relatively short storage stability, i.e., even though the compositions are stored in moisture- and airtight containers after compounding and prior to use, the polysulfide polymer compositions "cure-up" and lose their utility as adhesives, caulking compounds and the like. Because of the relatively short storage stability, these moisture activated polysulfide polymer compositions must be compounded shortly before use of the compositions. This often leads to inconvenience, undue expense and waste. For example, in case the composition is not used within this period of storage stability considerable waste may occur.

A process developed by E.F. Kutch (U.S. Patent 3,402,151; September 17, 1968; assigned to Thiokol Chemical Corporation) gives stable, curable liquid polythiopolymercaptan polymer containing zeolitic molecular sieves.

Examples 1 and 2: A one-part polysulfide polymer composition containing molecular sieves is compared to a control as to package stability.

	Parts By Weight	
	1	2
Polysulfide polymer [1]	100	100
Polyepoxide [2]	4	4
Calcium peroxide	8	8
Titanium dioxide	45	45
Toluene	8	8
Sodium hydroxide	0.5	0.5
Molecular Sieve [3]		4

[1] Polysulfide polymer having a molecular weight of about 4,000, a viscosity of about 40,000 centipoises, about 0.5% cross-linking, and essentially the structure

$$HS(C_2H_4-O-CH_2-O-C_2H_4-SS)_{23}C_2H_4-O-CH_2-O-C_2H_4-SH$$

[2] Bisphenol A-epichlorohydrin type epoxy resin.
[3] Dehydrated zeolitic material marketed as Linde 4A Molecular Sieves.

The ingredients were thoroughly mixed on a paint mill and then placed in an airtight lead lined toothpaste-type tube with a screw cap. The thus filled tubes are stored at 120°F. Periodically some of the material is extruded onto test panels from each of the tubes. A material is considered to have lost its package stability when the material is no longer extrudable. The composition of Example 1 is no longer package-stable at 39 days whereas the composition of Example 2 is still package-stable at 150 days.

Examples 3 through 6: Tests on different compositions prepared and tested as in Examples 1 and 2 are summarized as follows, in parts by weight.

	3	4	5	6
Parts by Weight:				
Polysulfide polymer [1]	100	100	100	100
Polyepoxide [2]	8	8	8	8
Barium peroxide	8	8	8	8
Titanium dioxide	45	45	45	5
Tris(dimethylaminomethyl)phenol			3	3
Ferric chloride	0.2	0.2	0.2	0.2
Molecular Sieve [3]		5		4
Package Stability (days):				
Storage at 75° F	125	>602	92	421
Storage at 120° F	92	>602	92	421

[1] Polysulfide polymer of Examples 1-2.
[2] An essentially anhydrous solution of 55±1% by weight of solid polyepoxide resin having a melting point of 125-135° C. and an epoxide equivalent value of 2,000-2,500 in a 1:1 solvent mixture of methyl isobutyl ketone and toluene.
[3] Dehydrated zeolitic material marketed as Linde 5A Molecular Sieves.

Examples 7 and 8: Compositions of the following compositions expressed in parts by weight were compounded in a room having a relative humidity of 8% at a temperature of 78°F. according to the procedure of Examples 1 and 2. The compositions were then tested for package-stability as in the previous examples.

	7	8
Parts of Weight:		
Polysulfide polymer [1]	100	100
Chlorinated biphenyl solvent	40	40
Titanium dioxide	15	15
Calcium carbonate	45	45
Molecular Sieve [2]		4
Package Stability (days): Storage at 75° C.	10	42

[1] Polysulfide polymer containing —NCO pendant groups and made according to the procedure of copending application S.N. 310,925, wherein 1 mole of an uncrosslinked hydroxyl terminated polysulfide polymer having a molecular weight of 2,000 is reacted with two moles of toluene diisocyanate.

[2] Dehydrated crystalline zeolitic molecular sieves marketed as Linde 4A Molecular Sieves.

COMPOSITION WITH ENCAPSULATED CURING AGENT

A process developed by A.S. Kidwell and N.R. Migdol (U.S. Patent 3,505,254; April 7, 1970; assigned to Interchemical Corporation) results in one-package curable polysulfide sealing composition comprising a liquid polyalkylene polysulfide having dispersed therein friable capsules of a polymeric film former encapsulating particles of a curing agent for the polysulfide. Curing of the sealing composition is accomplished by applying pressure to break the friable capsule thus liberating the curing agent.

The capsules are prepared using the fluidized bed technique and apparatus described in U.S. Patents 2,648,609 and 3,117,027. When a friable core is used, the core particles are the fluidized bed in the apparatus and a slurry of the curing agent particles and a binder for adhering the particles to the core in a suitable solvent are applied to the core as a spray. Then, the procedure is repeated with the core carrying the bound particles of curing agent as the fluidized bed and the polymeric film former which is to be the capsule being applied in solvent as the spray.

The particles of curing agent may be any solid curing agent used for polysulfides. While the peroxides are preferred as they are with the conventional two-package systems, any of the following curing agents may be used: ZnO, PbO, MgO, CaO, BaO, FeO, Fe_2O_3, CoO, CuO, ZnO_2, PbO_2, CaO_2, MnO_2, TeO_2, SeO_2, As_2O_3, Sb_2O_3, Sb_2O_5, SnO_2, Pb_3O_4, Na_2CrO_4, K_2CrO_4, $Na_2Cr_2O_7$, $K_2Cr_2O_7$, $(NH_4)_2Cr_2O_7$, $NaClO_4$, $KClO_4$, $Ba(ClO_4)_2$, $Na_2B_4O_7$, NH_4NO_2.

The polymeric film former is a friable material which is insoluble in the liquid polysulfide. Some suitable polymers having these properties are lignin sulfonates such as calcium magnesium lignin sulfonate, maleic modified phenol-formaldehyde, rosin modified phenol-formaldehydes, pentaerythritol esters of hydrogenated rosin such as Pentalyn H and copolymers of alpha-methyl styrene and vinyl toluene.

Example: Using the apparatus shown in U.S. Patent 3,117,027, 600 g. of pumice particles having an average particle diameter of 200 to 300 microns are coated with a slurry of

3,600 grams lead dioxide, 1,200 grams Chlorowax 70 (chlorinated paraffin wax having a 70% chlorine content by weight) and 3,600 grams toluene. The pumice particles are maintained as the fluidized bed upon which the slurry is sprayed. The fluidized bed is maintained by hot air at a temperature of about 140° to 160°F. During the coating, the toluene evaporates leaving a coating on the pumice particles of lead dioxide bound to the pumice by the Chlorowax 70. The coated particles are removed from the apparatus and passed through a 16 mesh screen to remove any agglomerates.

Then, using the same apparatus, the coated pumice cores are next encapsulated by forming a fluidized bed of the coated cores and spraying a slurry of 540 g. of calcium magnesium lignin sulfonate, 60 g. glycerine and 600 g. of water. The fluidized bed is maintained by hot air at a temperature of about 165° to 185°F. During the process, the water evaporates leaving a capsule predominantly of calcium magnesium lignin sulfonate around the coated core. The resulting capsules have a diameter in the range of 300 to 600 microns.

13 parts by weight of the above capsules are dispersed in a conventional polysulfide composition of 100 parts Thiokol LP-2, 30 parts carbon black, 5 parts dibutyl phthalate plasticizer, 2 parts silica, 20 parts calcium carbonate and 5 parts Durez 10694 (thermosetting phenol-formaldehyde resin). The resulting composition is very stable. After over 6 months storage at room temperature, no change is noted and the composition is still stable. After storage at 130°F. for over a month, the composition remained stable with no detrimental change.

To use and cure the composition, it is passed to an apparatus for mechanically crushing the capsules such as a two-roll mill. The composition is then applied, for example, as a joint sealant between window glass and an automobile frame. The composition cures to a tough rubbery material having all of the desirable properties of cured polysulfide compositions in which the lead dioxide is mixed with the polysulfide just prior to curing.

The stability of the composition may be even further increased by using a clay such as an aluminum silica clay in combination with the calcium magnesium lignin sulfonate in the capsule composition, up to 2 parts of the clay being used for each part of the sulfonate.

COMPANY INDEX

The company names listed below are given exactly as they appear in the patents, despite name changes, mergers and acquisitions which have, at times, resulted in the revision of a company name.

INVENTOR INDEX

U.S. PATENT NUMBER INDEX

NOTICE

Nothing contained in this Review shall be
construed to constitute a permission or recom-
mendation to practice any invention covered
by any patent without a license from the patent
owners. Further, neither the author nor the
publisher assumes any liability with respect to
the use of, or for damages resulting from the
use of, any information, apparatus, method or
process described in this Review.

ABS RESIN MANUFACTURE 1970

by C. Placek

Chemical Processing Review No. 46

ABS (acrylonitrile-butadiene-styrene) resins make up one of the most rapidly growing segments of the polymer industry. There are 39 producers of ABS throughout the world. U.S. consumption in 1969 was about 500 million pounds, with a total world production of about 1,200 million pounds. It is conceivable that U.S. production alone could reach 1,000 million pounds by 1975. This is a market to seek out for new and enlarged business ventures. This book is designed with that in mind.

This book offers detailed practical process information based on the patent literature for manufacture of ABS resins. The Table of Contents below indicates the type of information provided by this comprehensive survey.

1. Straight ABS Materials
   - Basic Process
   - Molded Vulcanized Mixture
   - Thermoplastic Molding Composition
   - Thermoplastic Casting Composition
   - Ternary Polymer Composition
   - Polycomponent Blends
   - Nonblooming ABS Composition
   - Graft Copolymers
   - Acrylonitrile Styrene (AS) Graft on ABS Copolymer
   - AS with Polybutadiene
   - ABS with Butadiene Styrene
   - AS Polymerized with Polybutadiene
   - Interpolymer with Rubbery Polybutadiene
   - ABS Polyblends
   - ABS Polymerized in Presence of Polystyrene
2. ABS Modified with Acrylic Derivatives
   - ABS with Acrylate Type Rubber
   - Acrylic Acid Esters and ABS
   - Blends with Poly (Methyl Methacrylate)
   - Rubbery Methacrylic Acid Terpolymer
   - Acrylate and Vinyl Alkyl Ether with ABS
   - Methyl Methacrylate with ABS
   - Use of n-tert-Butylacrylamide
3. ABS from Alpha-Methylstyrene (AMS)
   - Polymerization of Acrylonitrile/AMS
   - AMS/Acrylonitrile Blend with AMS/Diolefin
   - ABS with AMS/Acrylonitrile Copolymer
   - Interpolymer of Rubber Acrylonitrile and tert-Alkylstyrene
   - AMS and Methacrylate in ABS Composition
4. Miscellaneous Modifiers
   - Ethylenically Unsaturated Dicarboxylic Acid Anhydride
   - Use of Vinyl Chloride
   - Use of Divinyl Benzene
   - Copolymerization with Silane Diols
   - Epsilon-Caprolactam/ABS Blend
   - Sulfone/ABS Blend
   - Amine Modifiers
   - Itaconate Compounds
   - Carbonyls, Hydroxyls and Carboxyls
   - Ethylene/Vinylene Carbonate
   - Polycarbonate/ABS Blend
   - Dibutyl Fumarate
   - Mineral Oil
   - Hevea Rubber
5. Modification of Properties
   - Flame Resistance
   - Light and Temperature Stable
   - Antistatic
   - Improved Adhesion
6. Process Variations
   - Process Sequence
   - Cell Products
   - Monomer Addition to Latex
   - Elimination of Coagulation Problems
   - Specific Initiators
   - Two Stage Process
   - Biphase Plastics
   - Stable Aqueous Suspension
   - Coagulating ABS Latex
   - Continuous Monomer Addition
   - Stable Polymerization Solutions
   - Suspending Agents
   - Polymerizable Solvent Adhesive Composition
   - Incremental Addition of Styrene
   - Peroxide-Cured Blends
   - Solvents to Control Polymerization
   - Shock Resistant Polymers
   - Polymerization Along a Flow Path
   - Decomposition of Peroxy Oxygen

233 pages

$35

SOUNDPROOF BUILDING MATERIALS 1970

by M. Ranney

U.S. sales of noise abatement equipment are rising at an annual rate of 15 to 25%. This can be attributed in large measure to an increasing awareness of the health menace of excess noise and to the attendant interest in noise pollution abatement and also to some law enforcement action. The Walsh-Healey Act, which took effect in 1969, now regulates noise in factories for companies with government contracts in excess of $1,000. These factors signify a rapidly growing market.

The current market for noise and vibration control, a relatively untapped market, is placed at $170 million in annual sales, and this figure is expected to double within five years. Building materials for shelter construction are considered one of the areas with the most promising growth potential. A number of processes have been patented, largely relating to the use of fibrous glass, mineral wool and fiberboard; but gypsum, perlite, ceramics, and most recently, plastics have been studied in detail for this use. This book summarizes pertinent U.S. patent literature through March 1970 relating to soundproofing processes and techniques for this increasingly important industry.

The following shortened Table of Contents will provide an indication of the scope and content of this volume. The book's usefulness as a tool for making the most of this new market will become apparent from a study of the contents where the data contained is indicated by the numbers of processes included by the number in ( ).

1. Fibrous Glass and Mineral Wool
   - Fibrous Glass (14)
   - Surface Coating Techniques for Fibrous Glass (5)
   - Mineral Wool (10)
   - Flake Glass Panels (1)
2. Fiberboard
   - Perforation Techniques (6)
   - Panel Design (6)
   - Surface Coatings (3)
   - Miscellaneous (5)
3. Gypsum—Ceramic—Perlite
   - Gypsum (4)
   - Ceramic (3)
   - Perlite, Vermiculite and Silicate (5)
   - Miscellaneous (3)
4. Plastic Products
   - Polystyrene (3)
   - Polyurethane (4)
   - Miscellaneous (8)
5. General Processes
   - Flexible Partitions (3)
   - General Construction (7)

217 pages

$35

MEAT PRODUCT MANUFACTURE 1970

by E. Karmas

Food Processing Review No. 14

This most recent addition to the Food Processing Review series is concerned with the latest technology in the field of preparing packaged meats in both ready-to-cook and ready-to-eat forms. The information contained in this volume has been derived primarily from the U.S. patent literature since 1960, which provides you with comprehensive process information. This information has been divided into two sections—general processing methods, and specific products—to insure the greatest usefulness of the available material.

Meat has usually been eaten only at mealtimes because of the amount of time necessary for its preparation. However, with the aid of modern technology, meat is becoming increasingly available as both a snack food and a convenience item. Much of this book pertains to just these new types of meat foods.

The Table of Contents below indicates the many additional areas covered in this survey. The numbers in ( ) indicate the number of processes described under that heading.

A. General Processing

1. Curing Methods and Ingredients
   Accelerated Curing and Color Stability (13)
   Dietetic Curing Compositions (3)

2. Increased Water Binding and Yield
   Phosphates, Hydroxides and Other Chemicals (7)
   Various Yield Increasing Agents (5)

3. Improved Curing Formulations
   Nitrite Stability (6)
   Handling and Stability of Ascorbates (2)
   Homogeneity of Curing Compositions (7)

4. Integral Meats
   Mechanical Methods and Apparatus (5)
   Binding Agents (5)

5. Smoking
   Production of Smoke or Smoke Flavor Concentrates (5)
   Smoking Methods and Apparatus (4)

6. Thermal Processing and Sterilization
   Methods and Apparatus (2)
   Commercial Sterilization (2)
   Thermal Treatment (2)

7. Miscellaneous Processing Methods
   Auxiliary Curing (2)
   Product Releasing Agents (3)
   Meat Conditioning (2)

B. Products

8. Bacon Production
   Bacon Curing (5)
   Bacon Processing (3)
   Bacon Cooking (2)

9. Patty Type Products
   Fresh Patties (4)
   Dehydrated Patties with Binder (3)

10. Dehydrated Convenience and Snack Products
    Meat Tidbits (5)
    Pork Rind Snacks (3)

11. Modified and Novel Products
    Conventional Products (2)
    Filled and Shaped Products (2)
    Liquid or Powdered Meat Protein (2)
    Shelf-Stability with Antioxidants and other Additives (3)

273 pages

$35

EUROPEAN FOOD MARKET RESEARCH SOURCES 1970

by Noyes Data S.A.

This book is designed to serve as a guide to the mounting volume of statistical and market intelligence material currently available on the European food industry. It will enable the researcher to pinpoint, with the minimum of delay, those publications most likely to be of assistance, as well as indicate a wide range of sources which might not previously have been consulted.

The book contains an international section, listing sources with a world or European coverage and summarizing their contents. The book also deals with sixteen countries of Western Europe. In each case, references are classified under the following headings: (A) Government statistics and reports, (B) Other statistics and reports, (C) Trade associations, (D) Food trade journals, (E) Other newspapers and periodicals, (F) Directories, (G) Advertising statistics, (H) Bank reviews.

Although principally concentrating on the food processing industry, many sources have been included which also contain information on related sectors, such as fresh foods or self-service distribution. Entries are accompanied, where appropriate, by a synopsis of their contents and the name and address of the issuing authority. Since many statistics are collected and never published, these addresses will often be the key to further important sources of primary data.

European Food Market Research Sources' has been compiled with a view to providing a balanced and comprehensive selection of essential sources. It will prove an invaluable aid to those undertaking research into the food industries of Europe and a useful addition to the reference sections of commercial libraries.

This book was prepared by our subsidiary (Noyes Data S.A.) with offices in Zurich and London. On-the-spot coverage by our experienced European editors brings you a valuable, up-to-date marketing guide to this important industry.

111 pages

$19

PHARMACEUTICAL AND COSMETIC FIRMS U.S.A.

This publication describes the 600 leading pharmaceutical and cosmetic firms in the United States with as much information as possible to assist you in your sales, market research, acquisition, divestiture, or employment efforts. Includes (1) ethical, (2) proprietary, (3) veterinary, (4) private formula, (5) cosmetics, and (6) toiletries firms.

Contains this information (where available) for United States firms:

Name, Address, and Telephone Number
Ownership
Annual Sales Figure
Number of Employees
Names of Executives
Subsidiaries and Affiliates
Plant Locations
Products

This guide puts much information that is difficult to obtain, at your fingertips. Considerable effort was undertaken to include only those firms that have some significance. The pharmaceutical and cosmetic industry is difficult to assess because of the large number of very small companies; these companies have not been included. This enables you to concentrate your efforts on the larger firms.

This book has two indexes—a subsidiary and division index and a zip code index. The zip code index is an extremely important sales tool. This index lists the companies in numerical order by zip code, thereby providing you with a) easy-to-use, invaluable, geographical index.

Pharmaceutical industry sales in the United States are over $6 billion, and cosmetics sales are over $4 billion. These are two large industries, that are growing rapidly.

This guide can help you:

Concentrate on the big buyers
Prepare market reports
Increase sales effectiveness
Research potential acquisitions and divestitures
Employment and personnel guide

212 pages

$20

EDIBLE OILS AND FATS 1969
by Dr. N. E. Bednarcyk
Food Processing Review No. 5

This book describes in detail 225 recent process developments.

Shortenings: Fluid, Plastic, Miscellaneous; Margarine and Spreads; Margarine Oils, Highly Nutritional Oil Blends, Antispattering Agents, Fluid and Whipped Margarines, Flavor, Color, and Texture Modifications, Low Calorie Spreads: Salad Oils, Mayonnaise and Emulsified Dressings; Crystallization Inhibitors, Emulsified Dressings, Flavored Salad Oils, Low Calorie Dressings: Frying and Cooking Oils; Equipment, Breakdown Inhibitors, Antispattering Additives, Other Additives: Hard Butters; Preparation by Fractional Crystallization, Preparation by Ester Exchange, Miscellaneous: Oil Processing; Antioxidants and Stabilizers; Emulsifiers and Emulsions; Mixed Ester Emulsifiers, Dried Emulsion, Miscellaneous: Peanut Butter and Spreads: Chocolate Products; Indexes. Illustrations. 404 pages. $35

CONFECTIONARY PRODUCTS MANUFACTURING PROCESSES 1969
by M. Gutterson
Food Processing Review No. 6

This book is of technological significance in that it details over 200 processes for producing confections, based on the U.S. patent literature since 1960.

Based solely on new technology, this book offers substantial manufacturing information relating to this field. The wide scope of detailed data can be seen by the chapter headings indicated below:

Candy
Chocolate Products
Whipped Products
Icings
Gels
Coatings and Glazes
Gums and Stabilizers
Egg Products
Marshmallows and Meringues
Puddings
Frozen Confections
Chewing Gum
Other Confections
Indexes

Illustrations. 321 pages. $35

ALCOHOLIC MALT BEVERAGES 1969
by M. Gutcho
Food Processing Review No. 7

The traditional brewing process is a batch operation, costly and time consuming. There would be economic advantages to improved continuous processes which would require less capital investment for plant and equipment, give savings in labor, better use of raw materials, shorter processing time, and a more uniform product.

Detailed descriptive process information is found in this review, based on 157 U.S. Patents in the brewing field, issued since 1960. The 157 processes are organized in 7 chapters which tend to follow the steps in the brewing process.

Contents: Malting, Wort, Hops, Fermentation, Freeze Concentration and Reconstitution of Beer, Chillproofing, Preservation against Microbiological Spoilage, Foam, Indexes. Illustrations. 333 pages. $35

DEHYDRATION PROCESSES FOR CONVENIENCE FOODS 1969
by R. Noyes
Food Processing Review No. 2

Describes 236 up-to-date dehydration processes for producing specific foods. Most detailed body of information ever published.

The detailed, descriptive process information in this book is based on 236 U.S. patents in the food dehydration field—issued between January 1960 and May 1968. This book serves a double purpose in that it supplies detailed technical information, and can be used as a guide to the U.S. patent literature on dehydration of foods. By indicating only information that is significant, and eliminating much of the legal jargon in the patents; this book then becomes an advanced commercially oriented review of food dehydration processes.

Dry Milk Products, Cheese and Yoghurt, Eggs, Fruit and Vegetable Juices, Fruits, Potatoes, Vegetables, Coffee, Tea, Miscellaneous. Many illustrations. 367 pages. $35

PROTEIN FOOD SUPPLEMENTS 1969
by R. Noyes
Food Processing Review No. 3

The 126 Processes in this book are organized in 8 chapters by raw material source including the important newer processes for producing protein by fermentation of hydrocarbons. Another chapter on textured foods describes in detail a number of processes for producing these products that simulate meat. Indexes by company, inventors and patent number help in providing easily obtainable information.

This book is based upon the patent literature and serves a double purpose in that it supplies detailed technical information and can be used as a guide to the U.S. Patent literature on processes to obtain protein materials.

Contents: Hydrocarbon Fermentation, Fish-Based Protein, Soybeans, Cottonseed, Other Oilseeds and Legumes, Wheat and Gluten, Milk-Based Protein, Textured Foods, Miscellaneous, Indexes. Many illustrations. 412 pages. $35

SOLUBLE TEA PRODUCTION PROCESSES 1970
by Dr. N. Pintauro
Food Processing Review No. 11

This book describes production processes for producing soluble tea and offers a wealth of detailed practical information based primarily on the U.S. patent literature. Describes 73 specific processes in this field with substantial background information. The Table of Contents is listed below. The numbers in ( ) indicate the number of processes in that category.

Withering and Rolling (4)
Fermentation, Firing and Sorting (8)
Extraction (13)
Recovery of Aroma (10)
Tannin-Caffeine Precipitate (Cream) (15)
Filtration and Concentration (8)
Dehydration Process (6)
Agglomeration and Aromatization (9)

Illustrations. 183 pages. $35

FRESH MEAT PROCESSING 1970
by Dr. E. Karmas
Food Processing Review No. 12

This Food Processing Review, deals with 106 detailed processes covering essential developments in the fresh meat processing industry since 1960. The book provides a well-organized tour through the field; the processes included are well researched and presented as an easy-to-use guide to what is being done in this vital field today.

The material has been divided into two parts; processes for enhancing palatability, and preservation processes. The numbers in ( ) after each heading indicate the number of processes for each entry.

A. Palatability: Tenderness (33), Flavor and Tenderness (8), Flavoring (12), Color (13), Integral Texture (6). B. Preservation: Moisture Retention (9), Antimicrobial Treatment (10), Ionizing Radiation (7), Other Methods of Preservation (8). 236 pages. $35

217 pages. $35

SOLUBLE COFFEE MANUFACTURING PROCESSES 1969
by Dr. N. Pintauro
Food Processing Review No. 8

This book describes significant manufacturing processes for producing soluble coffee, and offers a wealth of detailed practical information based primarily on the U.S. patent literature. Describes 114 specific processes in this field with substantial background information.

Introduction: Roasting, Extraction, Filtration and Concentration, Recovery of Aromatic Volatiles, Spray Drying and Other Dehydration Processes, Freeze Drying Processes, Aromatization of Soluble Coffee Powder, Agglomeration Techniques for Soluble Coffee, Decaffeinated Soluble Coffee, Packaging of Soluble Coffee. Illustrations, Indexes. 254 pages. $35

SNACKS AND FRIED PRODUCTS 1969
by Dr. A. Lachmann
Food Processing Review No. 4

The sales of snack foods in the U.S. may reach the two billion dollar mark in 1969. Many companies are actively working on new snack foods or on improved processes. The patent literature on french fried potatoes, potato chips, corn chips and other crisps is continually growing and it is the purpose of this book to present this literature in easy readable form.

French fried potatoes and their methods of production are described in the second chapter. The next chapter deals with potato chips, still the most popular product of the snack food industry. The U.S. market for potato chips is estimated to be approximately 600 million dollars in 1969. In Chapter Four the processes for corn chips are covered; in Chapter Five, apple crisps. Chapter Six describes processes for expanded chips and some specialty items; and the last chapter deals with batter mixes. Many illustrations. 181 pages. $35

BAKED GOODS PRODUCTION PROCESSES 1969
by M. Gutterson
Food Processing Review No. 9

This book describes 201 recent processes for the production of baked goods. Based on the patent literature, it offers an up-to-date comprehensive publication of manufacturing processes.

There is a substantial amount of information in this book relating to the use of various chemicals and related additives.

Contents: Bread, Yeast Leavened Products, Chemically Leavened Products, Leavening Agents, Non-Leavened Products, Air Leavened Products, Refrigerated Doughs, Emulsifiers and Dough Improvers, Miscellaneous, Indexes. Illustrations. 353 pages. $35

MODERN BREAKFAST CEREAL PROCESSES 1970
by R. Daniels
Food Processing Review No. 13

Describes in detail production processes and equipment for the manufacture of modern breakfast cereals. These include both ready-to-eat and quick-cooking products.

Offers detailed practical information for the manufacture and production of these cereal products based on the U.S. patent literature. 61 processes included. Abbreviated Table of Contents follows.

Dough Cooking and Extrusion Processes
Treatment Prior To Puffing
Puffing Processes
Processes For Whole Cereal Grains
Cereal Shaping Processes
Sugarcoating Process
Fruit Incorporation and Nutritional Enrichment
Quick Cooking Cereal Products

217 pages. $35

ANIMAL FEEDS 1970
by M. Gutcho
Food Processing Review No. 10

This is a significant work, based on the patent literature, that shows you how to prepare a wide variety of modern feed products from numerous sources. It discusses the use of many chemicals, additives, and supplements used for animal feeds.

The 278 processes covered in this book are listed in the Table of Contents below. The numbers in ( ) indicate the number of processes described under that heading.

Introduction; Forage and Fodder (32), Fats and Oils (21), Molasses and Flavoring (15), Estrogens as Growth Stimulators (17), Antibiotics as Anabolic Stimulators (33), Antioxidants in Feeds (6), Minerals and Vitamins (13), Growth-Promoting Chemical Additives (24), Poultry Feeds (40), Ruminant Feeds (23), Feed for Swine (10), Pet and Other Feeds (19), Feed Products from Industrial Waste and By-Products (25). Indexes. 350 pages. $35

4

FREEZE DRYING OF FOODS AND BIOLOGICALS 1968
by R. Noyes
Food Processing Review No. 1

The detailed, descriptive process information in this book is based on 105 U.S. patents in the freeze drying field—issued between January 1960 and May 1968. Serves a double purpose in that it supplies detailed technical information, and can be used as a guide to the U.S. patent literature on freeze drying.

Higher costs of freeze drying are due to high capital costs for equipment, as well as higher operating costs due to high energy consumption and more limited output. This book contains many cost-cutting ideas.

Since the products obtained by freeze drying are considerably superior in most cases than those produced by other processes, considerable research and development work has been carried out attempting to lower costs. This book will give you the latest developments and recent advances. Numerous illustrations. 313 pages. $35

Noyes Data Corporation is the world's leading publisher of books relating to actual production techniques for producing specific products. They are down-to-earth practical books; theoretical considerations are only mentioned where pertinent to a basic understanding of the subject.

SYNTHETIC PERFUMERY MATERIALS 1970
by M. Gutcho
Chemical Processing Review No. 45

This Review shows you how to produce synthetic perfumery materials. It contains a valuable odor index.

The 152 U.S. patents included in this book are distributed among the 11 areas as shown below:

From Terpenic Materials (28)
Alcohols (11)
Esters (18)
Ethers (19)
Aldehydes (10)
Ketones (18)
Lactones, Pyrones, Substituted Phenols and Quinones (12)
Other Structures (7)
Naphthalene and Indene Derivatives (17)
Compounds with Scent of Ambergris or Irone (9)
Product Application (13)

273 pages. $35

ION EXCHANGE RESINS 1970
by C. Placek
Chemical Processing Review No. 44

This report on ion exchange resins provides detailed information on 126 U.S. patents issued since 1960 concerning the composition and manufacture of ion exchange materials. This Review, by its organization, also provides a guide to these ion exchange resins by grouping them according to physical form, behavior characteristics, etc.

1. Anion Exchange Resins
2. Cation Exchange Resins
3. Resins For Removing Metals
4. Resins Having Mixed Properties
5. Unconventional Materials
6. Specific Use Resins
7. Process Emphasis
8. Properties of Ion Exchange
9. Ion Exchange Membranes
10. Emphasis on Shapes

329 pages. $35

PHOTOCHEMICAL PROCESSES 1969
by B. Albertson
Chemical Processing Review No. 36

The purpose of this Review is to provide an up-to-date description of industrial photochemical technology as recorded in the U.S. Patent literature since 1960. Describes in detail 210 photochemical production processes.

Photochemistry, long known as a selective powerful tool, has excited the imagination of chemical engineers as a method of manufacturing various chemicals. From the economic point of view, photochemistry has the following advantages: 1. Many materials cannot be obtained in any other way. 2. Specificity of effects cannot be achieved by other methods. 3. Less need for high pressures and temperatures can lower costs.

Introduction, Photohalogenation, Photonitrosation, Organic Photochemical Reactions, Inorganic Photochemical Reactions, Photopolymerization, Indexes. Illustrations. 185 pages. $35

ALKALI METAL PHOSPHATES 1969
by Dr. M. W. Ranney
Chemical Processing Review No. 34

The 97 processes described are based on U.S. patents issued since 1960 and offer a comprehensive treatment of up-to-date technical information.

Contents: Orthophosphates, Metaphosphates, Pyrophosphates, Tripolyphosphates, Phosphites/Hypophosphites. Numerous illustrations. 344 pages. $35

AMINES, NITRILES AND ISOCYANATES PROCESSES AND PRODUCTS 1969
by M. Sittig
Chemical Process Review No. 31

Material covered includes: Manufacture of Amines, Manufacture of Mono-Nitriles, Acrylonitrile Derivatives, Isocyanate Manufacture, Future Trends. 62 illustrations. 201 pages. $35

ALCOHOLS, POLYOLS, AND PHENOLS MANUFACTURE AND DERIVATIVES 1968
by M. Sittig
Chemical Process Review No. 23

Contents: Introduction, Manufacture of Alcohols, Manufacture of Glycols, Manufacture of Polyols, Manufacture of Phenol, Reactions of Alcohols, Reactions of Glycols and Polyols, Phenol Derivatives. 71 illustrations. 201 pages. $35

PHTHALOCYANINE TECHNOLOGY 1970
by Y. L. Meltzer
Chemical Processing Review No. 42

Advances in phthalocyanine technology have been truly explosive during the past few years. New phthalocyanine products, processes and applications have poured forth from industrial, governmental and academic laboratories at a rapid pace. These advances in technology have made themselves felt in the market place and in government programs, and have contributed to corporate sales and profits. At the same time, however, competition has become more intense in the phthalocyanine field making it imperative to keep up with the latest technological advances.

Examines recent developments in phthalocyanine technology as reflected in U.S. patents and other literature. The first 23 chapters discuss up-to-date manufacturing processes for phthalocyanine products and dyes. Chapters 24 through 31 discuss unusual new applications for phthalocyanines. 390 pages. $35

RADIATION CHEMICAL PROCESSING 1969
by R. Whiting
Chemical Processing Review No. 41

A number of radiation induced chemical processes are already operating commercially. The radiation processing of chemicals has been growing at an annual rate of about 25% per year. Currently, $100 to 150 million worth of irradiated products are produced in the United States per year, however, it has been forecast that by 1980, the value of products receiving radiation treatment will be close to $1,000 million per year.

This book surveys the radiation processing field and is based on the U.S. patent literature since 1960. Over 250 separate processes are described in detail in the chemical, polymer, rubber, petroleum, textile and other fields. Contents: Polyolefins, Other Polymers, Elastomers, Hydrocarbons, Organic Chemicals, Inorganic and Organo-Metallic Compounds, Other Processes, Indexes. 377 pages. $35

CHLORINE AND CAUSTIC SODA MANUFACTURE RECENT DEVELOPMENTS 1969
by Dr. R. Powell
Chemical Processing Review No. 33

Contents:
General Considerations
Preparation of Brine for Electrolysis
Brine Electrolysis in Diaphragm Cells
Brine Electrolysis in Mercury Cells
Combination Mercury Cell and Amalgam Fuel Cell System
Recovery of Mercury from Brine
Production of Caustic Soda in the Decomposer Unit
Coordinated Operation of Diaphragm and Mercury Cells
Platinum-Coated Titanium Anodes
Electrolysis of Sea Water
Electrolysis of Hydrogen Chloride
Catalytic Oxidation of Chloride
Cracking of Hydrogen Chloride
Nitrosyl-Chloride Route
Cooling, Drying and Purification of Chlorine
Purification of Caustic Soda

Numerous illustrations. 265 pages. $35

CITRIC ACID PRODUCTION PROCESSES 1969
by R. Noyes
Chemical Processing Review No. 37

Detailed descriptions of production processes for citric acid, based on the patent literature. The Table of Contents is indicated below:

Processing, Iron Impurities, Other Microorganisms, Recovery and Purification, Other Processes. Indexes. 157 pages. $24

PRACTICAL DETERGENT MANUFACTURE 1968
by M. Sittig
Chemical Process Review No. 27

Contents: Manufacture of Branched-Chain Olefins, Linear Alpha Olefins, and Linear Paraffins; Competitive Routes to Straight-Chain Alcohols, Alkylaromatics and other Detergent Raw Materials; Sulfation and Sulfonation Processes; Detergent Formulation. 76 illustrations. 212 pages. $35

POLYACETAL RESINS, ALDEHYDES, AND KETONES 1968
by M. Sittig
Chemical Process Review No. 26

Polyformaldehyde or polyoxymethylene is assuming increased importance.

Contents: Introduction, Production of Aldehydes, Production of Ketones, Reactions of Aldehydes, Reactions of Ketones, Aldehyde Polymers, Future Trends. 81 illustrations. 237 pages. $35

MONOSODIUM GLUTAMATE AND GLUTAMIC ACID 1968
by Dr. R. Powell
Chemical Process Review No. 25

This technical Review with numerous flow diagrams and equipment designs, describes in detail processes for manufacture of MSG.

U.S. demand in 1970 is expected to be 80 million pounds. World production of MSG was 240 million pounds in 1967. Numerous illustrations. 256 pages. $35

HYDRAZINE MANUFACTURING PROCESSES 1968
by Dr. R. Powell
Chemical Process Review No. 28

General Considerations, Raschig Process, Bayer Process, Processes Based on Urea, Electrical Glow Discharge Processes, Chemo-Nuclear Processes, Direct Production of Anhydrous Hydrazine, Other Processes, Recovery, Concentration, Stabilization. Numerous illustrations. 252 pages. $35

ELECTRODEPOSITION AND RADIATION CURING OF COATINGS 1970
by Dr. M. W. Ranney

The advantages of electrodeposition: pinhole free coating, automated operation, elimination of fire hazards and air pollution problems, low operating costs and fast throughput make this an attractive method.

Radiation curing will also reach significant commercial status, due to its advantages of rapid curing, elimination of ovens, and use of solvent-free vehicles. **1. Electrodeposition—Techniques**—Control of Bath Stability, Electrical Parameters, Miscellaneous Feed Control, **2. Electrodeposition—Formulation**—Acid-Containing Resins, Cathode Deposition, Miscellaneous Emulsions, Nonaqueous Systems, **3. Electrodeposition—General**—Metal Treatment, Fillers and Pigments, Miscellaneous, **4. Radiation Curing—Formulations**—Polymeric Vehicles, Wood Impregnation **5. Radiation Curing—Equipment** 170 pages. **$35**

CELLULAR PLASTICS RECENT DEVELOPMENTS 1970
by K. Johnson

Recent accomplishments in the production of cellular plastics includes among the 190 total processes, 71 polyurethane processes and a number of techniques for polyolefin and polystyrene foam. Included information can be seen in the abbreviated Table of Contents where the numbers in ( ) indicate how many processes are covered in each area. Also contains company, author, and patent number indexes which will help you make the greatest use of the information.

Polyolefins (15)
Polyvinyl Chloride (15)
Polystyrene (22)
Rubber (15)
Polyurethanes (71)
Polyesters and Epoxides (15)
Urea-Formaldehyde and Phenolic Resins (9)
Other Cellular Products (27)

280 pages. **$35**

SYNTHETIC FIBERS FROM PETROLEUM 1967
by M. Sittig
Chemical Process Review No. 1

Production processes for the four major synthetic fibers: nylon, polyesters, acrylics, and polyolefins. Contains a wealth of processing data with particular emphasis on process conditions. A technical evaluation of the synthetic fiber industry. 116 illustrations. 275 pages. **$35**

STEREO-RUBBER AND OTHER ELASTOMER PROCESSES 1967
by M. Sittig
Chemical Process Review No. 3

Synthesis of new polymers with rubber-like elasticity is a vital growing industry. Modification of molecular architecture of elastomers for specific properties is subject of worldwide research. Review summarizes state of the art in elastomer industry; new routes to monomers and polymerization techniques. 215 pages. 77 illustrations. **$35**

FLAME RETARDANT POLYMERS 1970
by M. Ranney

Summarizes selected process technology for the use of fire retardant imparting additives and reactive intermediates for major polymeric plastic materials, with particular emphasis on recent technology in the areas of polyesters, polystyrene and polyurethane foam. There are 144 separate processes included, all based on the U.S. patent literature.

An abbreviated Table of Contents is listed. The numbers in parentheses indicate the number of processes included for each entry.

Polyethylene and
  Polypropylene (15)
Polystyrene (19)
Polyurethanes (50)
Polyesters (13)
Other Polymer Systems (26)
General Utility Additives (21)

263 pages. **$35**

PLASTIC FILMS FROM PETROLEUM RAW MATERIALS 1967
by M. Sittig
Chemical Process Review No. 6

This Review concerns production of synthetic polymer films. Contents: Raw Materials, Polymer Manufacture, Film Production Processes, Film Composition, Film Treating Processes, Future Trends. 118 illustrations. 276 pages. **$35**

POLYOLEFIN PROCESSES 1967
by M. Sittig
Chemical Process Review No. 2

Manufacturing techniques and processes for polyolefin resins. Includes: Current Status of Various Polyolefins, Processes Using Non-Metallic Catalysts, Metal Oxide Catalysts, Metal Alkyl-Reducible Metal Halide Catalysts; Polymer After-Treatment, and Fabrication. 96 illustrations. 234 pages. **$35**

PLASTICIZER EVALUATION AND PERFORMANCE 1967
by I. Mellan

This book will help you evaluate: 1. A new plasticizer with a known resin; 2. A known plasticizer with an unknown resin; 3. A new plasticizer with a new resin.

The 110 tables of performance data, and the standard methods of testing described in this book, provide data from which one can estimate what a particular plasticizer will do in a specific resin.

The plasticizers were chosen for inclusion, primarily by their greatest demand in the industry. Chapters: 1. Introduction; 2. Testing; 3. Comparative Performance Data; 4. Performance Data of Individual Plasticizers; 5. Brand Names and Their Manufacturers; 6. Abbreviations for Coding Plasticizers; 7. Chemical Names of Plasticizers and Their Brand Names; 8. Brand Names of Plasticizers and Their Chemical Names; 9. Bibliography. 178 pages **$20**

SYNTHETIC LEATHER FROM PETROLEUM 1969
by M. Sittig
Chemical Process Review No. 29

The newer synthetic leathers offer permeability to water vapor and air, as does natural leather. In addition, the newer products offer good tear strength and softness. It is expected that these newer synthetic leathers will make substantial inroads in shoe and other markets.

These modern synthetic leathers consist, in general, of nonwoven mats of such fibers as polyesters impregnated with such binder resins as polyurethanes. The fiber-binder combination is then rendered porous by one of a number of different processes. Introduction: Manufacture of Fiber-Forming Polymers, Types of Structural Fibers Used, Mat Formation, Mat Treatment, Binder Polymer Formulation, Other Ingredients, Fiber-Binder Combination, Consolidation, Coating, Introduction of Porosity, Surface Modification, Product Properties, Future Trends. 84 illustrations. 214 pages. **$35**

COMPATIBILITY AND SOLUBILITY 1968
by I. Mellan

Normally, it requires laborious testing to determine compatibility of polymers, resins, elastomers, plasticizers, and solvents. Predictions made without testing or literature searching, are usually unreliable. We all know of the many commercial products that were failures, when compatibility and solubility considerations were ignored.

This book helps you evaluate proper materials by the use of 224 tables. The tables were originally published by manufacturers of various materials, and reproduced in this book, to give you a reference work to this important subject. The tables are organized in three sections—polymers and resins, plasticizers and esters, and solvents. The tables in each section indicate the solubility and compatibility of the particular material with a wide range of other materials. 304 pages **$20**

CARBON BLACK TECHNOLOGY RECENT DEVELOPMENTS 1968
by Dr. R. Powell
Chemical Process Review No. 21

Introduction; Feedstocks, Channel Blacks, Furnace Blacks, Thermal Blacks, Acetylene Blacks, High Structure Carbon Blacks, Low Structure Carbon Blacks, Unconventional Processes, Carbon Black Pelletizing, Other Finishing Treatments. Numerous illustrations. 242 pages. **$35**

AROMATICS MANUFACTURE AND DERIVATIVES 1968
by M. Sittig
Chemical Process Review No. 17

Contents: Introduction, Production of Aromatics, Separation of Aromatics, Purification of Aromatics, Reactions giving Hydrocarbon Products, Other Reactions, Phenol Production, Styrene Manufacture and Derivatives, Future Trends. 73 illustrations. 232 pages. **$35**

SUGAR ESTERS 1968
by Research Corporation

Already approved for use in foods in a number of countries, use of sugar esters in foods in the United States awaits FDA clearance.

Sugar esters are an important new raw material for the food industry.

Contains papers presented at California Symposium 1967. 134 pages. **$15**

HYDROGEN PEROXIDE MANUFACTURE 1968
by Dr. R. Powell
Chemical Process Review No. 20

This Review is concerned with chemical processes for manufacturing hydrogen peroxide. Major emphasis is placed on the anthraquinone processes. General Considerations, The Anthraquinone Process, Other Processes, Purification, Concentration, Stabilization. Numerous illustrations. 221 pages. **$35**

CATALYSTS AND CATALYTIC PROCESSES 1967
by M Sittig
Chemical Process Review No. 7

Contents: Hydrocarbon Conversion Processes, Hydrocarbon Polymerization Processes, Hydrocarbon Oxidation Processes, Future Trends. 109 illustrations. 303 pages. **$35**

INDUSTRIAL GASES MANUFACTURE AND APPLICATIONS 1967
by M. Sittig
Chemical Process Review No. 4

This book discusses conventional cryogenic air separation and purification techniques in considerable detail.

This book also discusses newer techniques such as adsorption using molecular sieves, and permeation using various membrane materials. 313 pages. 103 illustrations. **$35**

TITANIUM DIOXIDE AND TITANIUM TETRACHLORIDE 1968
by Dr. R. Powell
Chemical Process Review No. 18

Describes titanium dioxide processing steps (a) chlorination of rutile ore, (b) separation of sublimated solids, (c) purification of crude titanium tetrachloride, and (d) conversion of titanium tetrachloride to titanium dioxide. Numerous illustrations. 306 pages. **$35**

ELECTRO-ORGANIC CHEMICAL PROCESSING 1968
by Dr. C. Mantell
Chemical Process Review No. 14

This volume has been written from the viewpoint of the chemical engineer, with emphasis on plant processes, operating data, and plant design. The commercial successes in this field have been attained by the chemical engineering approach. 186 pages. **$35**

SOIL RESISTANT TEXTILES 1970
by Dr. M. W. Ranney
Textile Processing Review No. 5

Ideal soil release finishes must be capable of releasing stains readily and preventing redeposition of soil during laundering. Treatments should render manmade fibers and durable press reactants less attractive to oily stains and should be more easily wetted.

This report summarizes the developments in soil retardant and soil release finishes in both the carpet industry and in textile manufacture. It includes the newest technology associated with the use of acrylates and fluorochemical treatments. The numbers in ( ) following each treating agent indicate the number of processes covered for that particular compound.

Introduction: Metal Oxides and Salts For Carpet Treatment (15), Acrylic and Vinyl Polymers (10), Silicones (6), Fluorochemical Compounds (72), General Treatments (14). 216 pages. $35

WATERPROOFING TEXTILES 1970
by Dr. M. W. Ranney
Textile Processing Review No. 4

This Textile Processing Review summarizes the technology of water resistant treatments for textiles and fabrics as described in the U.S. patent literature since the early 1950's. 246 waterproofing processes are included—64 relate to use of fluorochemicals.

The numbers in ( ) after each entry in the Table of Contents, where the treatment processes are organized by the agent used, indicates the number of production processes for each agent.

Production Processes for Waterproofing Textiles using the following Agents: Metal Salts and Wax-Containing Formulations (44), Silicones and Alkyl Polysiloxanes (53), Organofunctional Silicones and Fluorosilanes (20), Acrylics (8), Nitrogen Containing Compounds (30), Fluorochemical Compounds (64), Elastomer, Vinyl, Polyolefin Vapor Permeable Fabrics (27), Miscellaneous Treatments (9). 353 pages. $35

FLAME RETARDANT TEXTILES 1970
by Dr. M. W. Ranney
Textile Processing Review No. 3

Describes 177 commercial processes to produce flame retardant textiles and fabrics.

Most activity is based on chemical modification of cellulose through hydroxyl groups. Use of phosphoric acid, urea-phosphates, and other phosphorylating agents all confer flame retardant properties to cellulose. A significant portion of this book is devoted to the latest in application of phosphorous containing materials.

Numbers in ( ) indicate the number of processes described.

Ammonium Salts, Borates (12); Antimony, Titanium Metal Oxides (25); Amine-Phosphorus Products (21); Aziridines, APO, APS (21); Methylol-Phosphorus Polymers, THPC (27); Phosphonitrilic Chlorides (9); Trialkyl Phosphates and Phosphonates (26); Silicones Isocyanates, Miscellaneous (10) Nylon, Acrylics (18). 373 pages. $35

THERMOELECTRIC MATERIALS 1970
by M. Sittig
Electronics Materials Review No. 7

Manufacturing processes for:

Alkali Mn Tellurides, Sn and Bi Selenides and Tellurides, Bi Telluride, Boron Matrices, BP, Cd-Sn Arsenide, Ce Sulfide Matrices, Ce Titanate, CsCl Melt, Cr and Co Silicide, GeTe, Au-Ni Alloys, Au-Ag Combinations, Graphite, In-Ga Arsenide, Pb-Pb Dioxide, PbTe, Mn-Ge Telluride, Mn Silicide, HgTe, Organics, Rhodium Arsenide Antimonide, Ruthenium Arsenide Antimonide, Sm Sulfide, Si-C Matrices, Si-Ge Alloys, Ag-Sb Telluride, Ag Selenide, Sr Titanate, Thallium Telluride, Tin Oxide, Tin Telluride, Ti-Va Ceramics, Uranium Dioxide, Uranium Monosulfide, Uranium Nitrate, Zn Antimonide, Zr in Zirconia Matrix.

73 illustrations. 235 pages. $35

PHOTOCONDUCTIVE MATERIALS 1970
by M. Sittig
Electronics Materials Review No. 8

Photoconductors have a number of important applications, such as; television camera tubes, solar cells, photoelectric cells, solid state light amplifiers, electrophotographic copying processes.

They offer the promise of a proliferation of applications in the near future such as new forms of computer circuitry, optical carrier communications systems using laser light sources, and in color copying processes. There are 86 processes relevant to the manufacture of these materials. They cover the three areas indicated in the abridged Table of Contents.

Electrophotographic Materials
Vidicon Tube Materials
Photocell Materials

288 pages. $35

MAGNETIC MATERIALS 1970
by M. Sittig
Electronics Materials Review No. 9

This latest review in the Electronics Materials series surveys the empirical state of the art with regard to magnetic materials as revealed primarily in recent U.S. patents.

In arranging the sections of the text which number three as can be seen from the Table of Contents below, the author has listed the various compositions alphabetically according to principal component. Ferrites containing both magnesium and manganese may be covered twice—once as magnesium manganese ferrites where the magnesium is the principal constituent, and once as manganese magnesium ferrites where manganese is the principal constituent.

Simple Magnetic Oxide Materials, Complex Magnetic Oxide Materials, Magnetic Metallic Materials. 286 pages. $35

CREASEPROOFING TEXTILES 1970
by Dr. M. W. Ranney
Textile Processing Review No. 2

Summarizes with detailed process information relating to textile creaseproofing agents used to obtain wash and wear, or permanent press fabrics. Over 300,000 words, describes 343 processes in this field. Shows you chemical agents used, and processes by which they are applied.

Dimethylolethylene Urea and Related Compounds, Aldehyde-Urea Condensates, Uron Resins, Aminoplasts—Catalyst Performance, Melamine Derivatives, Triazones, Carbamates, Other Nitrogen-Containing Compounds, Phosphorus-Amino Compounds—Aziridines, Aminoplast-Thermoplastic Resin Compositions, General Processing Techniques and Formulations, Aldehydes, Acetals, Epoxies, Epihalohydrins, Sulfones, Sulfonium Salts, Cross-Linking Agents, Miscellaneous, Polymeric Coatings—Rubber, Vinyl, Silicones, Radiation Curing, Wool, Nylon, and Others. Indexes. 460 pages. $35

DYEING OF SYNTHETIC FIBERS 1969
by C. Whiting
Textile Processing Review No. 1

The dyeing of synthetic fibers continues to be a challenge. Successful methods however, have been developed. Chapters 2 through 7 of this Review are divided into two sections. The first presents various new dyes and dyeing processes which are applicable to synthetic fibers. The second section concerns many auxiliary products available to aid in production of an acceptably dyed product. These include dye improving agents, leveling and retarding agents, agents used in after treatment to achieve optimum fastness properties.

Introduction: Dyeing Polyolefin and Polypropylene Fibers, Dyeing Polyamide Fibers, Dyeing Polyester Fibers, Dyeing Acrylic Fibers, Dyeing Hydrophobic Fibers, Dyeing Glass Fibers, Dyeing Miscellaneous Fibers, Dyes Applicable to More than One Type of Fiber, Dyeing Fiber Blends. 257 pages. $35

ELECTROLUMINESCENT MATERIALS 1970
by M. Sittig
Electronics Materials Review No. 6

The phenomenon of electroluminescence offers a variety of proven and potential applications such as illuminated display panels and color TV tubes. Much research and development is being put into the evolution of this new breed of luminescent materials.

This Electronics Materials Review is concerned with manufacturing processes for producing electroluminescent materials. The numbers in ( ) indicate the number of processes given for producing each of the indicated electroluminescent materials. This book, based on the latest U.S. patent literature, serves as a guide to a productive new industry.

Introduction, Cathode Ray Tube Phosphors (20), Color Television Phosphors (25), Electroluminescent Lamp Materials (20), Electroluminescent Diode Materials (14), Future Trends. 306 pages. $35

DOPING AND SEMICONDUCTOR JUNCTION FORMATION 1970
by M. Sittig
Electronics Materials Review No. 5

Donor or acceptor impurities may be added to the liquid phase from which semiconductor crystals are grown or they may later be deposited on the crystal surface and diffused inward. In either case the addition of such impurities is called doping. Where there is formed a transition between a P-type and an N-type material as a consequence of one of these operations, a semiconductor junction is formed. This book describes 115 processes for doping and semiconductor junction formation.

Production of Alloyed Junctions, Diffusion Processes, Melt Grown Junctions, Doping During Melting, Simultaneous Dopant and Substrate Deposition, Spark Doping Processes, Doping by Particle Bombardment, Hydrothermally Grown Junctions, Doping Epitaxially Grown Layers, Future Trends. Indexes. 132 Illustrations. 318 pages. $35

PRODUCING FILMS OF ELECTRONIC MATERIALS 1970
by M. Sittig
Electronics Materials Review No. 4

Films of varying thickness have a multitude of applications in the fabrication of modern electronic devices. This book describes numerous processes for film production as indicated in the Contents below. The numbers in ( ) indicate the number of processes described under that heading.

Cathode Sputtering (10)
Vacuum Evaporation (18)
Gas Plating (2)
Electroplating (6)
Electroless Deposition (6)
Film Deposition From Reacting Vapors (32)
Explosive Evaporation (1)
Squeezing on Optical Flats (1)
Electron Deposition (2)
Ion Beam Deposition (1)
Reaction & Deposition From Solution (5)
Oxide Films by Reaction at Solid Surfaces (7)
Carrier Transport in Vapor Phase (6)
97 Illustrations. 295 pages. $35

SEMICONDUCTOR CRYSTAL MANUFACTURE 1969
by M. Sittig
Electronics Materials Review No. 3

1. Introduction
2. Verneuil Method
3. Other Fusion Processes
4. Spark Discharge Process
5. Single Crystal Pulling From a Crucible
6. Single Crystal Production by Zone Melting
7. Other Melt Processes
8. Dendritic Crystal Production From Melts
9. Dendritic Crystal Production From Vapor
10. Single Crystals by Sintering
11. Melting Under Pressure
12. Crystals by Deposition From Vapor Phase
13. Crystals by Deposition From Vapor Phase Reaction
14. Hydrothermal Growth
15. Crystal Growth in Solution Melts
16. Shaped Crystals
17. Single Crystal Film
18. Future Trends
   Indexes

106 illustrations. 303 pages. $35

PURE CHEMICAL ELEMENTS FOR SEMICONDUCTORS 1969
by M. Sittig
Electronics Materials Review No. 2

Detailed manufacturing techniques for producing pure chemical elements for semiconductors. Antimony, Arsenic, Bismuth, Boron, Gallium, Germanium, Indium, Phosphorus, Selenium, Silicon, Tellurium, Thallium, Zinc, Future Trends. 112 illustrations. 335 pages. $35

MANUFACTURE OF SEMI-CONDUCTOR COMPOUNDS 1969
by M. Sittig
Electronics Materials Review No. 1

Manufacturing processes to produce semiconductor compounds based on Aluminum, Bismuth, Boron, Cadmium, Calcium, Godolinium, Gallium, Indium, Iron, Lead, Magnesium, Silicon, Silver, Thallium, Titanium, Vanadium, Zinc, and Organics. 106 illustrations. 326 pages. $35

COSMETIC FILMS 1970
by M. Gutcho

62 processes with methods of forming cosmetic films.

Powders
 Absorptive Powders
 Surgical Powders
 Cosmetic Powders
 Compressed Powders
 Aerosol Powders
Creams and Lotions
Nail Preparations
 Pearlescent Fingernail Coatings
 Improve Durability of Nail Coatings
 Polish Removers
 Nail Strengtheners
 Mending Nails
 Artificial Nails
Lipsticks
 Lipstick Bases
 Improved Dyes and Pigments
 Combining Masses of Varying Hardness
 Shaped Crystals
 Multicolored Lipsticks
 Liquid Lip Rouge

143 pages. $20

ANTIOBESITY DRUG MANUFACTURE 1970
by Dr. B. Idson
Chemical Processing Review No. 43

This book discusses 162 processes in the field of anorectic drug production technology. The numbers in ( ) in the Table of Contents show the number of preparations discussed for each heading.

N-Alkyl Amines (2), Substituted Aminopropanes—Arylaminopropanes (17), Substituted Aminopropanes (2), Substituted Aminopropanes—Bicyclic Aminopropane (1), Substituted Aminopropanes—1-Heterocyclic Oxoaminopropanes (1), Substituted Aminoalkenes (2), Aralkyl Amines (22), Aralkyl Hydrocarbon (1), Ring Substituted Heterocyclic Compounds (20), Nitrogen Substituted Heterocyclic Compounds (43), Bicyclic Compounds (2), Azacyclic Compounds (5), Miscellaneous Functional Substitutions (24), Alkaloids (1), Antibiotic (1), Resins (3), Steroids (6), Amphetamine Salts (2), Compositions (8), Appendix. 193 pages. $35

NUCLEOTIDES AND NUCLEOSIDES 1970
by S. Gutcho

Basic biological research is an area of great scientific progress today. Introduction of new equipment and new technologies has increased the pace of biochemical research with resultant major forward strides into the search for the basic compounds of life.

The Table of Contents indicates the data offered by this book. 159 processes are described. The numbers in ( ) indicate the number of preparation processes covered.

Organic Synthesis of Nucleotides in General (16), Synthesis of Specific Nucleotides (18), Fermentation Procedures for Nucleotides in General (8), Fermentation Procedures for Specific Nucleotides (31), Enzymatic Digestion of Nucleic Acids (3), Nucleotides Coenzymes (26), Cyclic Nucleotides (2), Dinucleoside Phosphates (9), Purification Techniques (13), General Procedures (23), Nucleotides as Flavor Enhancers (11).

200 pages. $35

HAIR PREPARATIONS 1969
by A. Williams

Up-to-date technology of hair care products.

The rapidly rising demand for hair care preparations has resulted in a steady stream of new products for both the consumer and the professional beauty shop operator over the past few years. Active research and new product formulation have been carried out by a number of firms. This book discusses a significant portion of these later developments.

This book provides a detailed technological summary of recent developments based on 138 U.S. patents, since 1960, covering all aspects of hair preparations for the head, beard, eyelashes, and eyebrows.

Introduction, Dyeing, Bleaching, Waving, Setting, Shampoos-Rinses, Grooming-Tonics, Shaving Assistants, Other, Indexes. 208 pages. $35

DENTIFRICES 1970
by T. Jefopoulos

The most common dental therapeutic agents in use today are the fluorides, ammonium compounds, penicillin and antienzyme agents. The decay preventing value of the fluorides has made them the most widely used anticaries material. Dental formulations also contain ingredients which contribute to the polishing and cleaning functions of many dentifrices. This book offers a guide to the information available in the U.S. Patent literature with regard to both therapeutic and cosmetic agents in dentifrices.

The numbers in ( ) after each entry in the Table of Contents indicate the number of processes or formulations covered.

Introduction
Cleaning Agents (17)
Polishing Agents (14)
Prophylactic Compositions (27)
Fluorides (30)
Dentifrices For Dentists (6)
Other Dentifrices (30)
Improved Dentifrice Processing (2)

191 pages. $35

SUSTAINED RELEASE PHARMACEUTICALS 1969
by A. Williams

This book is based on the U.S. patent literature, and presents substantial technical information for production of these products.

TABLETS-PILLS
 Cellulosic Coatings
 Lipid Coating
 Gels-Gums
 Specific Medicaments
 Tablet Design
 Miscellaneous
CAPSULES
 General Coatings
 Particle Size Control
 Capsule Design & Manufacture
INJECTABLES
 Antibiotics
 Miscellaneous
MISCELLANEOUS RELEASE PREPARATIONS
 Suppositories
 Powders
 General

Indexes. 273 pages. $35

TETRACYCLINE MANUFAC-TURING PROCESSES 1969
(2 Volumes)

CTC, Oxytetracycline, TC, DMTC, and DMCTC, 2N-Derivatives, Position 4 Derivatives, 6-Methylene Derivatives, Anhydrotetracyclines, 7-and/or 9-Derivatives, 11a-Halo Derivatives, 5a, 11a-Dehydrotetracyclines, 12a-Deoxy Derivatives, 12a-Derivatives, Epimers, Mechanism Study intermediates. Indexes. 931 pages. (2 Volumes) $45

VITAMIN B₁₂ MANUFACTURE 1969
by R. Noyes
Chemical Process Review No. 40

Vitamin B₁₂ active substances are important therapeutic products for treatment of pernicious anemia. Also used for treatment of various other human ailments, and as a veterinary growth factor. This book offers various methods of producing vitamin B₁₂ active substances. 327 pages. $35

VITAMIN E MANUFACTURE 1969
by T. Rubel
Chemical Processing Review No. 39

This review relates the known methods for the preparation of tocopherols from natural products or by synthetic means, and conversion of non-alpha tocopherols.

Introduction; Tocopherols From Deodorizer Sludge, Conversions to Alpha Tocopherol, Synthesis of Tocopherols, Miscellaneous Related Processes, Indexes. 114 pages. $24

FIBRINOLYTIC ENZYME MANUFACTURE 1969
by T. Rubel
Chemical Processing Review No. 38

This Review presents some methods for the production and purification of fibrinolytic agents, their precursors and activators. Plasminogen and Fibrinolysin, Urokinase, Streptokinase and Streptodornase, Other Fibrinolytic Enzymes. Indexes. 139 pages. $24

EUROPEAN KNITWEAR AND HOSIERY MARKET REPORT 1970
by Noyes Data S.A.

17 countries included. Trend of activity established. Size structure of domestic market determined. Information on yarn consumption, knitwear and hosiery production, foreign trade, number of companies and employees engaged in manufacture. Comprehensive selection of statistical material.

In addition, the report contains a list of leading knitwear and hosiery companies. In all, the names and addresses of over 1,000 major European manufacturers are supplied, an indispensable aid to sales promotion and research projects.

This report presents a coherent picture of the progress being made by the knitting industry throughout Western Europe. It will help assess the potential of existing markets and plan the penetration of new ones. 166 pages. $35

TEXTILE GUIDE TO EUROPE 1970
by Noyes Data S.A.

This is the fourth in our series of directories dealing with key industries of Western Europe.

The detailed company information provided falls into two categories. In Part 1, entries are listed alphabetically under country and are described wherever possible, with the following data: full name and address; principal executives; product range; domestic and foreign subsidiaries and affiliates; plant location; latest sales figures and numbers of employees. In Part 2, the arrangement of companies is again by country but this time classification is according to the type of textile products manufactured.

Contains no advertising material. Layout is simple and easy to follow. Entries can be located rapidly and information extracted without reference to complicated lists of explanatory symbols and abbreviations. A French, German and Spanish vocabulary of key words has been included. 220 pages. $20

PESTICIDE PRODUCTION PROCESSES 1967
by M. Sittig
Chemical Process Review No. 5

Production processes for these pesticides are described: Olefin-based Insecticides; Olefin-based Fungicides; Olefin-based Herbicides; Diolefin-based Insecticides; Diolefin-based Fungicides; Diolefin-based Herbicides; Aromatic-based Insecticides; Aromatic-based Fungicides; Aromatic-based Herbicides. 48 flow diagrams. 200 pages. $35

UREA PROCESS TECHNOLOGY 1968
by Dr. R. Powell
Chemical Process Review No. 13

The information in this book is up-to-date, and is based on information made available since 1960. The information in this volume includes a huge variety of numerical data.

General Considerations, Processes, Low Biuret Processes, Product Finishing and Controlled Release Fertilizer. Illustrations. 324 pages. $35

AMMONIA AND SYNTHESIS GAS 1967
by R. Noyes
Chemical Process Monograph No. 26

Outlines tehnology and economics pertaining to ammonia production, with many flow diagrams and tables. Various methods of hydrogen and synthesis gas preparation, shift conversion techniques, various carbon dioxide removal systems, final purification, compression, and ammonia synthesis. Bibliography. Many illustrations. 175 pages. $20

POTASH AND POTASSIUM FERTILIZERS 1966
by R. Noyes
Chemical Process Monograph No. 15

Introduction: Markets, Future Projections; Geology and Geography; Underground Deposits of Soluble Minerals; Brines; Other Sources of Potash; Flotation; Crystallization; Other Refining Processes; Potassium Sulfate; Potassium Nitrate; Potassium Phosphates; Final Product. 210 pages. $15

FERTILIZER DEVELOPMENTS AND TRENDS 1968
by A. V. Slack

This book analyzes the opportunities for improving fertilizer technology based on current world-wide trends in research and development.

Chemistry of fertilizers and current manufacturing methods are discussed only to the extent necessary for adequately relating new departures to standard practice. The major emphasis is in departures that have been introduced in the past five or six years, and new developments that will assume major importance in the future.

Contents: Research and Development, Ammonia, Ammonium Nitrate, Urea, Ammonium Sulfate, Slow Release Nitrogen, Other Nitrogen Fertilizers, Phosphoric Acid, Ammonium Phosphate, Nitric Phosphates, Superphosphate, Thermal and Miscellaneous Phosphate Processes, Potassium Fertilizers, Fluid Fertilizers, Bulk Blending, Minor Nutrients. 98 illustrations. 406 pages. $35

NITRIC ACID TECHNOLOGY RECENT DEVELOPMENTS 1969
by Dr. R. Powell
Chemical Process Review No. 30

Ammonia Oxidation Process, Wisconsin Thermal Process, Nitrogen Fixation by Shock Waves, Nitrogen Fixation in a Nuclear Reactor, Absorption of Nitrogen Oxides in Water, Concentration of Dilute Nitric Acid Solutions, Direct Production of Concentrated Nitric Acid, Purification of Nitric Acid, Stabilizers for Nitric Acid. Numerous illustrations. 245 pages. $35

PHOSPHORIC ACID BY THE WET PROCESS 1967
by R. Noyes
Chemical Process Review No. 9

This book is devoted to production of phosphoric acid by the wet process. Production technology is changing rapidly, with new single tank reactors, the advent of superphosphoric acid, new commercial processes for acidulation with hydrochloric and nitric acids, new semihydrate and anhydrite processes, etc. 282 pages. $35

NEW FERTILIZER MATERIALS 1968
by C.I.E.C.

This book is invaluable to fertilizer and chemical manufacturers who are looking to produce or market potentially useful new fertilizer materials.

Ureaform, Crotonylidenediurea (CDU), Isobutylidene Diurea, Superphosphoric Acid, Triple Superphosphate of 55% P_2O_5 Content, Ammonium Phosphates, Ammonium Polyphosphate, Nitrophosphates, Nitrate of Potash, Potassium Metaphosphate (Potassium Polyphosphate), Monopotassium Phosphate and Monoammonium Phosphate, Magnesium as a Plant Nutrient, Magnesium Phosphates, Magnesium Ammonium Phosphate and Related Compounds, Sulfur as a Fertilizer, Oxamide, Urea Nitrate, Urea Phosphate, The Hydrides of Phosphorus, Red Phosphorus, Fertilizer Application Equipment, Modern Use of Fertilizer. Numerous Illustrations. 430 pages. $35

AMMONIUM PHOSPHATES 1969
by Dr. M. W. Ranney
Chemical Process Review No. 35

This book describes recent processes for production of ammonium phosphates.

Introduction; Ammonium Orthophosphates, Diammonium Orthophosphates, Ammonium Polyphosphates, Metal Ammonium Phosphates, Ammonium Phosphate–Ammonium Nitrate Mixtures. Many illustrations. 278 pages. $35

CONTROLLED RELEASE FERTILIZERS 1967
by Dr. R. Powell
Chemical Process Review No. 15

This book offers you complete technical data on numerous processes and products in this field. The two major approaches are (a) compounds of low solubility, and (b) coated granules.

Introduction; Compounds of Low Solubility, Coated Granules, Prevention of Nitrogen Losses, Rapid-Release Fertilizer. 279 pages. $35

EUROPEAN AGRICULTURAL CHEMICALS SURVEY 1969
by S. A. Mann

This is a guide to the European Agricultural Chemicals Industry that will help you in sales, market research, development, planning, acquisitions and mergers. The pesticides considered in this survey include all plant or crop protection chemicals such as insecticides, herbicides, fungicides, etc.

There are three types of information presented for each country: 1. Discussion of the market structure. 2. Statistical material on production, consumption, imports, and exports. 3. The names and addresses of the leading pesticide producers in each country. This book contains a total of about 1,000 names and addresses of Europe's leading pesticide producers.

This book contains considerable statistical material, invaluable for market research studies. Much of the statistical material is broken down in considerable detail. 129 pages. $35

EUROPEAN MAN-MADE FIBER MARKET REPORT 1968
by S. A. Mann

This book will give you an overall picture of the man-made fiber situation in Western Europe. It is a valuable work that will help you in market research studies. Never before has so much statistical information on this subject been included between the covers of one publication.

There are numerous tables pertaining to production, consumption, exports and imports. 186 pages. $35

EUROPEAN CHEMICAL MARKET RESEARCH SOURCES 1969
by Noyes Data S.A.

This is a book which will prove an invaluable aid to those undertaking research into the rapidly expanding chemical markets of Western Europe. Designed to serve as a guide to the numerous sources of statistical and market information currently available on this industry, it will enable the researcher to pinpoint, with the minimum of delay, those publications most likely to be of assistance, as well as indicate a range of sources which might previously have remained unknown.

Each country is treated separately, sources being grouped under the following headings: Government statistics and reports, Other statistics and reports, Trade associations, Trade journals, Other newspapers, and magazines, Directories, Advertising statistics, Bank reviews. 102 pages. $19

EUROPEAN FOOD PROCESSING INDUSTRY 1968
by S. A. Mann

A guide to European Food Processing Industry that will help you in sales, market research, development, planning, acquisitions, and mergers in this large growth industry relating to processed and convenience foods.

For 18 countries, includes discussion, statistical material, and names and addresses of about 2,000 major food processers. 197 pages. $35

ELECTRONICS GUIDE TO EUROPE 1969
by Noyes Data S.A.

Electronics is currently one of the most dynamic sectors of European industry, with a growing number of participating companies manufacturing an ever expanding range of products.

It contains up-to-date company profiles of over 600 leading electronics manufacturers in 14 countries.

The comprehensive company information included in this guide will greatly assist in the planning of effective marketing operations. Wherever possible, entries are accompanied by the following key data arranged for easy comparison: Full name and address, Ownership, Principal executives, Product range, Domestic and overseas subsidiaries, Plant location, Latest sales figures, Number of employees.

This book will help you: Increase sales, Prepare market reports, Know company officials, Locate suppliers, Organize joint ventures, Make licensing arrangements. 150 pages. $20

FOOD GUIDE TO EUROPE 1969

This is a directory of the European Food Processing Industry that will help you in sales, market research, development, planning, acquisitions, and mergers. It is expected that the manufacture of processed foods in Europe will be a substantial growth industry.

Describes the 800 leading European food processing firms in 18 countries: Name and Address, Ownership, Plant Locations and Products, Local Subsidiaries and Affiliates, Foreign Subsidiaries and Affiliates, Principal Executives, Annual Sales, Number of Employees.

A valuable marketing guide. Only the 800 major food processing firms are included.

Will help you: Increase sales, Concentrate on the big buyers, Plan joint ventures, Make licensing arrangements, Know company officials, Prepare market research reports, Talk intelligently about the European food processing industry. 130 pages. $20

CHEMICAL GUIDE TO EUROPE 1968
Fourth Edition

Describes the 1,000 leading European chemical manufacturers in 18 countries. Gives name and address, ownership, plant locations, products, local and foreign subsidiaries and affiliates, principal executives, annual sales and number of employees of the firms tht have most to offer by way of sales contacts, licensing arrangements and joint ventures. 207 pages. $20

Books relating to European industry are published by a subsidiary, Noyes Data S.A., with offices in Zug, Switzerland and London, England.

STARCHES AND CORN SYRUPS 1970
by Dr. A. Lachmann

This report covers the field of starch production from many standpoints.

Wet milling is the primary method of starch production, therefore much of the material is concerned with this route to starch. Dry milling processes are also covered.

In addition, coverage of the current technological progress in hydrolyzing starches into dextrins, corn syrups and dextrose and starch fractionation into amylose is covered.

Contains 139 processes covering: The Manufacture of Starch, Treatment of Starch, Modified Starch, Pregelatinized Starch, Acid Hydrolysis of Starch to Sweeteners, Enzymatic Starch Hydrolyzing Enzymes, Starch Hydrolysates, Starch Hydrolysates Produced by Acid and Enzyme Treatments, Starch Fractionation. 275 pages. $35

CHEMICAL ZIP BOOK 1970
by H. Bennett

The usefulness of this book derives from its organization. It is divided into two sections. The first section is an alphabetical listing by name of the approximately 4,000 United States chemical firms giving, in addition to the correct name of the company, both its address and zip code. The second section is arranged numerically according to zip code with the companies once again listed alphabetically within their proper zip code numbers, thereby providing you with an easy-to-use, invaluable geographical index to the chemical industry of the United States.

Both the main office address of a company and any divisional addresses, where they differ from the corporate headquarters address, are supplied. The addresses provided by this book offer a valuable and easily used tool for locating the primary buying power in the chemical industry—a major factor in good sales planning. 134 pages. $15

FIRE RETARDANT BUILDING PRODUCTS AND COATINGS 1970
by Dr. M. W. Ranney

Imparting fire retardancy presents a many-sided challange to the building industry. The fire retardant chemicals must be relatively inexpensive because of the economics of the building trade. In addition the coatings must also be able to maintain the generally desirable characteristics of good building materials such as paintability and hygroscopicity.

These needs, which are present in the search for good fire retardants, are surveyed in this Review. The value of this report is indicated by the number of chemicals and processes covered for all the building materials here.

1. Wood Impregnation
2. Fiberboard
3. Ceiling Tile and Panel Construction
4. Asphaltic Products
5. Intumescent Coatings
6. General Coating Formulations
7. Adhesives

186 pages. $35

MICROENCAPSULATION TECHNOLOGY 1969
by Dr. M. W. Ranney

This book is based on U.S. patent technology in the microencapsulation field. Detailed descriptions and illustrations of processes and products are given. An abbreviated table of contents is outlined below:

Phase Separation Methods; Coacervation-Aqueous Phase Separation, Organic Phase Separation, Spray Drying, Miscellaneous: Interface Reactions—Polymerization; Dissolved Monomer Polymerization, Interfacial Polymerization, In-situ-Polymerization, Vapor Deposition: Physical Methods; Fluidized Bed, Electrostatic, Multi-Orifice Centrifugal, Vacuum Metallizing, Coating Fusible Material: Applications; Xerographic Toner, Light Sensitive-Photographic Materials, Heat Sensitive, Transfer Sheets, Dyes, Miscellaneous, Indexes. Illustrations. 275 pages. $35

ORGANIC CHEMICAL PROCESS ENCYCLOPEDIA
by M. Sittig
Second Edition 1969

Gives the key processing facts for instant reference for 711 industrial organic chemical processes—with 711 large flow diagrams.

This book offers a handy desktop reference to organic chemicals and their industrial processes. The primary purpose of this book is to present the significant processing conditions all in one place—saving you many hours of work trying to obtain specific numbers.

The format of the book is such that the key process facts can be picked out quickly—they are on the same place on every page. Process facts included are: equation; feed materials; coproducts; type of catalyst; phase; reactor type; solvent used; temperature; pressure; reaction time; heat required or heat evolved; product yield; product purity; materials of construction; major product uses; and reference. 712 pages—8½″ x 11″—hard cover. $35

SOLAR CELLS 1969
by Dr. M. W. Ranney

Semiconductors—Silicon; Ce Panel Fabrication Techniques Semiconductors—Cadmium, Gallium; Photoemission; Organic Photochemical Cells; Miscellaneous Applications. 85 illustrations. 271 pages. $35

FUEL CELLS RECENT DEVELOPMENTS 1969
by Dr. M. W. Ranney

This book describes in detail a wide range of recent developments in fuel cell technology, based on the latest U.S. Patents.

Fuels; Electrolytes; Membranes-Separators; Electrodes; Fuel Cell Construction; Biochemical and Thermocells; Regeneration and Reactivation. 325 pages. $35

INDUSTRIAL SOLVENTS HANDBOOK 1970
by I. Mellan

A handbook with complete, up-to-date, pertinent data regarding industrial solvents.

821 tables contain pertinent data concerning physical properties of solvents and degrees of solubility of materials in these solvents. Numerous graphs included giving a great deal of data concerning various parameters. Also includes phase diagrams for multi-component products.

The vast amount of information contained in this book is shown in the abbreviated Table of Contents in the next column. The numbers in ( ) after each entry indicate the number of tables.

Hydrocarbon Solvents (14); Halogenated Hydrocarbons (130); Nitroparaffins (5); Organic Sulfur Compounds (5); Monohydric Alcohols (122); Polyhydric Alcohols (150); Phenols (6); Aldehydes (10); Ketones (53); Glycol Ethers (79); Ketones (44); Acids (18); Amines (124); Esters (61). 478 pages. $25

WATER POLLUTION CONTROL AND SOLID WASTES DISPOSAL 1969
by M. Sittig
Chemical Processing Review No. 32

Describes specific water pollution processes, a substantial growth industry. Much of the technology necessary to effect water pollution control and solid wastes disposal has already been developed. In many cases these environmental problems could be resolved by intelligent process choice, and economic application of the proper control equipment.

This book in most cases covers processes and equipment which can be licensed or purchased.

Introduction: The Water Problem, The Solid Waste Problem, Types of Water Contaminants, Removing Specific Inorganic Water Contaminants, Removing Specific Organic Water Contaminants, Removing Specific Solid Contaminants from Water, Handling Liquid and Solid Radioactive Wastes, Sewage Disposal, Solid Waste Disposal Processes, Future Trends. 78 illustrations. 244 pages. $35

AIR POLLUTION CONTROL PROCESSES AND EQUIPMENT 1968
by M. Sittig
Chemical Process Review No. 24

It is to a great extent application of known apparatus, processes, and concepts that will enable us to control air pollution.

Most complete descriptions of processes and equipment are available in the patent literature. These then, in most cases, cover processes and equipment which can be licensed or purchased today to combat environmental control problems. Only through sound engineering principles in process design are we going to control air pollution at minimum cost. This book gives detailed process descriptions.

Contents: Sources of Air Pollution, Air Pollution Control Devices, Removing Specific Gases and Vapors from Air, Equipment for Removing Solids and Liquids from Air, Removing Specific Solids and Liquids from Air, Removing Automotive Exhaust Fumes from Air. Future Trends. 102 illustrations. 260 pages. $35

GUIDE TO CHEMICAL PLANT PLANNING 1969
by Dr. R. Lobstein

The assemblage of information needed for selecting materials and equipment, and for fitting them into the dimensional scheme, can present a major problem to the most experienced of engineers and architects. Endless perusal of catalogs, technical magazines, and handbooks often are not specifically concerned with the problem at hand becomes necessary. One purpose of this book is to make much of the searching effort unnecessary. The required information has been collected and condensed into clear, quickly available essential data.

Much of the book is intended as an aid in the selection of mechanical equipment. Many tables and diagrams in current usage by equipment and machinery suppliers are included.

Contents: Introduction, Piping, Pumps, Heat, Power Transmission, Mechanical Equipment, Electrical Engineering, Operational Requisites. 452 charts. 524 pages. $24

INORGANIC CHEMICAL AND METALLURGICAL PROCESS ENCYCLOPEDIA 1968
by M. Sittig

Today, there is more "glamour" in many inorganic chemicals and metals than there is in organic chemicals. For example, gallium arsenide, now important in the laser and semi-conductor field was of no significance until only recently. The same can be said for many other inorganic and metallurgical chemicals that are now assuming importance due to rapidly changing developments in electronics, lasers, integrated circuits, cryogenics, superconductivity, materials development, nuclear energy, and the space program.

This book is organized in an unusual format. There is one inorganic chemical or metallurgical process on each page. At the top of the page an equation or flow diagram is shown, and underneath a description of the process is given. 883 pages—8½″ x 11″—hard-cover. $35

POLYMETHYLBENZENES 1969
by H. W. Earhart

Presents: physical properties of PMB's and derivatives; chemistry of the PMB's; known and suggested end-uses for numerous PMB derivatives.

Compilation of information for: Benzene, Toluene, Xylene, Mesitylene, Pseudocumene, Hemimellitine, Durene, Isodurene, Prehnitene, Pentamethylbenzene, Hexamethylbenzene. 549 literature references. $20

MEMBRANES TECHNOLOGY AND ECONOMICS 1967
by Dr. R. Rickles

Chapters: Membranes Theory, Electrodialysis, Ultrafiltration, Dialysis and Diffusion Control, Membrane Filtration, Gas Permeation, Medical Applications, Preparation of Synthetic Polymeric Membranes. 197 pages. $24